Thick Film Technology

Fundamentals and Applications in Microelectronics

Thick Film Technology

Fundamentals and Applications in Microelectronics

JEREMY AGNEW

Vicon Instrument Company

HAYDEN BOOK COMPANY, INC.
Rochelle Park, New Jersey

To my wife, Sylvia

ISBN 0-8104-5803-9
Library of Congress Catalog Card Number 72-90079

Printed in the United States of America

1 2 3 4 5 6 7 8 9 PRINTING

73 74 75 76 77 78 YEAR

Preface

Although some thick film techniques have been around for many years, it is only recently that thick film has been accepted as the answer to many circuit fabrication problems. In the years to come, thick film circuitry will become more and more widespread as its many advantages are appreciated. It is the purpose of this book to introduce engineers and technicians to this emerging technology and to prepare them for the inevitable new developments that will occur.

A book such as this, which tries to cover the field of thick film microelectronics, could not have been written without the direct and indirect help of countless individuals. It has taken the efforts of many people working in this field to develop thick film electronics from a stop-gap method of circuit fabrication to its present level of sophistication. This book is based on the efforts of all the people working with thick film technology who have brought it to today's state of the art.

I would like to take this opportunity to thank the many people who helped me—through discussions, meetings, and practical advice—when I made my start in the field of thick film processing. I hope this volume will be helpful to others who follow.

Colorado Springs, Colorado JEREMY AGNEW

Contents

1 — The Hybrid Process

The electronic industry has come a long way since the audion tubes of Lee De Forest in the early 1900's. The addition of a grid structure to control the electron flow observed in vacuum diodes by Sir John Ambrose Fleming was to have a profound effect on our way of life. Radios and amplifiers became common and, by the 1940's, other strange electronic marvels, such as radar and computers, were in use. Then one of the most important discoveries of the twentieth century took place. Scientists at Bell Telephone Laboratories observed the solid state phenomena, leading to the invention of the point-contact transistor in 1948.

Immediately, electronics took a giant step forward. The increased reliability, smaller sizes, and lower power requirements of the transistor made products appear that were not possible before. The vacuum tube computer, which had been suffering from high failure rates and giant power requirements, was to benefit tremendously. Transistor radios that were no bigger than a pack of cigarettes appeared in shirt pockets all over the country.

The next leap forward occurred with Texas Instruments' announcement of the first integrated circuits in 1959. Whole circuits built on a single wafer of silicon could outperform the discrete versions, as well as be packaged in an even smaller space.

By the late 1960's terms such as LSI and MSI were becoming part of the engineer's vocabulary. Large Scale Integration (LSI) and Medium Scale Integration (MSI) promised whole systems on a single wafer. Now a whole computer could be fabricated on a single slice of silicon, instead of taking up a large air-conditioned room with rack after rack of vacuum tubes. After the introduction of integrated circuits, it appeared that every requirement could be met by these silicon wafers, with their impurities controlled in localized areas and selectively metallized interconnections, forming monolithic circuits to synthesize the desired function.

Upon closer examination, not every requirement can be met by these circuits. Totally integrated monolithic circuits are expensive to make. The tooling costs can run as high as \$40,000 per circuit. It would not be economical to produce fifty or a hundred circuits this way. The circuit, once it is designed, is expensive to change. Several months' work and high retooling costs are involved. Monolithic circuits do not perform well at very high frequencies, such as are in the microwave region. Monolithics are basically very low power devices. Eight transistors and miscellaneous resistors on a 100-mil-square die do not stand much dissipated power in the form of heat before being destroyed or changing characteristics. And yet a discrete circuit is not always the answer either. Smaller physical size, fewer connections, and lower power requirements are the more desirable characteristics.

This trend towards miniaturization, while still maintaining circuit flexibility, led to the hybrid integrated circuit. Hybrid circuits tend towards miniaturization by removing the large packages and carriers from discrete components and mounting the actual working parts of the component directly on a ceramic substrate.

The fabrication of hybrid integrated circuits relies on knowledge from several disciplines including chemistry, metallurgy, and physics. The support for the circuit is a thin wafer of ceramic. Conductive patterns are deposited onto this by screen-printing techniques using metallic pastes. This pattern takes the place of hand-wired interconnections. Resistors are deposited in the same way using a resistive paste similar to that used to form the element in a carbon potentiometer. Capacitors can be added in the circuit by printing layers of metal and dielectric. Transistors are added as bare chips of silicon, just as they come from the manufacturer's wafer. Connections to the transistors are made by bonding wires finer than a human hair to the base and emitter and then down to the metallized substrate. Finally the entire circuit is sealed into a tiny package or encapsulated in some manner to protect all the components. Because of the relatively thick films of paste screened onto the ceramic substrate, this type of hybrid circuit is called a thick film circuit.

The reliability of a hybrid circuit is greater than the reliability of a discrete circuit since the number of circuit interconnections is reduced. For example, a chip-and-wire transistor in a hybrid circuit has two interconnections, instead of the three found in a discrete circuit. A diode has one in a hybrid instead of two. The remaining electrical connection is made through the physical connection to the substrate. There are still the same number of electrical connections to the device in both cases, but the methods of making the connections are different.

The flexibility of a discrete circuit is maintained in hybrid form because pieces of the circuit can be added bit by bit as and where desired. The circuit can be changed easily because the tooling is not as complex as it is for a monolithic circuit. Thick film tooling costs can run as low as \$100. Because of the low tooling costs, thick film hybrids can be used for short production runs. Ten or a hundred circuits can be produced without the massive start-up costs and time delays of the monolithic circuits.

Hybrid integrated circuits are actually divided into two basic types—thick film hybrids and thin film hybrids. Thick film hybrid circuits use the techniques of screen printing to deposit select geometries of resistive and conductive pastes on a ceramic substrate. The patterns are then fired at high temperatures to form a hard film, approximately a mil thick. Thin film materials are vaporized in a high vacuum and deposited as a thin coating onto glass or very smooth ceramic substrates. Then patterns are selectively etched into the conductive and resistive coatings to form resistors with connecting conductive paths. Thin films are typically only several hundred Angstroms thick. From this point on the two technologies merge for active and passive device attachment and packaging.

There are various pros and cons to circuit manufacturing by thick film or by thin film techniques. Until relatively recently the trend was to use the thin film approach for hybrid circuits. Deposited thin films afforded far greater precision and smaller line widths than printed thick film resistors and conductors. However, in recent months thick films seem to be overtaking thin films. Newer precision pastes and better printing equipment have led to excellent results with thick film circuits. Some of the advantages are as follows:

1. A wider resistance range is obtainable with thick films. Thin film resistor materials can produce resistors with values of from about 10 ohms/square to 4,000 ohms/square. (See Chapter 2 for an explanation of the ohms-per-square concept.) This means that for high value resistors, such as 500 kilohms, the resistor would have to be at least 100 squares long. Large geometry resistors take up too much valuable real estate on the substrate and thus become impractical. Thick film pastes are available up to 10 megohms/square. These can be used to produce physically small, high-value resistors.
2. Thick film resistors have a better power dissipation capability than thin film resistors.
3. Turnaround time for thick film is very quick. Tooling is relatively low cost. Thick film tooling may typically run \$500 to \$1000. Thin film tooling typically costs around \$1000 to \$2000. Thus, the cost per circuit is usually higher for thin film—ordinarily one-and-one-half to two times more than for thick film.
4. Thick film is capable of higher production rates than thin film. The screening and firing process for thick films can crank out many millions of circuits. Thin films use more delicate photoresist and chemical etching techniques.

5. Initial equipment investment for thick film is much lower than for thin film. A typical, small thick film facility costs from \$25,000 to \$50,000, as opposed to \$40,000 to \$70,000 or higher for a similar thin film facility. A vacuum deposition chamber, along with the associated plumbing and pumps, costs from \$10,000 to \$20,000 just to start with.

On the other side, there are the following disadvantages:

1. Thin film techniques can produce finer line widths for greater circuit density. Thin film lines may be typically one mil, as opposed to five mils for thick films.
2. Thin films can produce closer initial tolerances than thick films. Typically, thin films may be ±5% as deposited, as compared to 10 to 20% for screened and fired thick films.
3. Noise levels and temperature coefficients of resistance (TCR) are typically lower for thin films. An average figure may be three to five times better than for thick film.

There is a place for both thick films and for thin films, depending on the application. Actually the circuits are not all that different. Only the resistor- and conductor-making parts of the process are different. Identical methods and materials are used in die and wire bonding, and final packaging. Thin film substrates have a smoother surface than do thick film substrates to provide a better deposition surface. These substrates are more expensive than the coarse alumina substrates used for thick films, which is one reason for the increased cost per circuit for thin films.

Despite the previously mentioned drawbacks of monolithic circuits, they should not be forgotten. For high-volume, low-cost circuits, monolithic technology is superior to that of the hybrid circuit. Once the tooling is fixed and paid off, circuits can be mass-produced extremely cheaply. Because of the piecemeal nature of hybrid circuit assembly, the costs will tend to level out after a certain production volume has been reached.

The following is a list of the steps in a typical thick film process:

1. Clean substrate.
2. Screen conductor pattern onto substrate.
3. Dry conductor pattern.
4. Fire conductor pattern.
5. Screen resistor pattern onto substrate.
6. Dry resistor pattern.
7. Repeat steps 5 and 6, if necessary, for different values of paste.
8. Fire resistor pattern.
9. Trim resistors to desired values.

Fig. 1-1. A screened and fired thick film resistor pattern. Substrate on the left is a 1″ x 1″ alumina substrate with palladium-silver conductors screened on it. Substrate on the right has resistors added.

10. Attach transistor dice to circuit by die bonding.
11. Attach passive components.
12. Wire bonding to form circuit interconnections.
13. Attach external leads if necessary.
14. Inspect and test.
15. Package the circuit.
16. Final testing.

Figure 1-1 shows a substrate with resistors and conductors fired on it as a test pattern.

The following chapters will cover the details of making thick film hybrid circuits—from artwork to the final packaging. The text does not attempt to discuss particular pieces of equipment or processes, but tries rather to make a general survey and comparison of thick film processes along with some general details of problems and procedures. Each thick film manufacturer has to make up his own particular process based on his equipment, his design, and his goals. There is still a great deal of "witchcraft" involved in circuit making that can only be learned by working with hybrids. Because of differences in equipment, personnel, and environment, a process that will work for one person in one location cannot necessarily be duplicated somewhere else by another manufacturer. Thus, in many cases, it is impossible to give a specific recommendation on a particular process or the minute details involved in carrying it through. For this reason the reader must bear with the text if it seems somewhat general at times. The only way to settle on a thick film manufacturing process is to experiment. Try the various techniques based on individual needs, and settle for the ones that work best.

Much of the equipment used in thick film work is purchased as a "custom" item. The variety of options offered by the manufacturers on equipment is quite wide and the ones selected depend on how the particular thick film process has evolved. The type of circuits actually produced will determine the equipment needed. Therefore it would not be suitable to discuss particular equipment at any great length other than to specify its basic operation. Specific manufacturers will all be happy to provide literature and assistance on their products. Another good source of state-of-the-art developments is current technical periodicals. The International Society for Hybrid Microelectronics (ISHM) is devoted exclusively to hybrid technology and provides a valuable meeting ground for people working in the hybrid field. The state-of-the-art technology in thick film work is changing all the time, but armed with the basic principles involved, the practicing engineer, technician, or production manager should be able to master any new developments that come along.

2 — Artwork and Layout

The starting point for a thick film hybrid circuit is generation of the artwork. The circuit has been designed and checked. It may have been breadboarded with discrete components or simulated on a computer. In some fashion the circuit design has been finalized, and now the task is to convert it into a working microcircuit.

The shape and size of the resistors, as well as their locations, have to be laid out onto the substrate. The pattern of interconnections has to be figured out and the location of transistors decided on. The layout of a hybrid circuit is no different than drawing up a blueprint for a machine design or layout of a printed circuit board—all of them are roadmaps for building a product. In hybrid circuits the only difference is that the components are not the normal resistors and capacitors associated with electronic circuitry.

Suggested guidelines for circuit layout and spacing are detailed at the end of this chapter for reference only. The actual dimensions given will change as the technology advances. As materials and processes are refined and equipment is designed to work to even closer tolerances, the dimensions given will shrink. However, some items will not change much. An average transistor chip is about 20 mils square and diodes about 15 mils square. This will probably not become much smaller in the future because of the physical limitations of making and handling a 2-mil-square transistor. Anyone who has dropped a transistor chip and has turned over pieces of dust and dirt with needlepoint tweezers trying to find it again will appreciate the problems involved. Similarly, when screening a thick film resistor, the mesh size of the screen becomes a limiting factor. It is quite easy to produce a thin film circuit line 2 mils wide by means of photographic techniques, but trying to accurately screen-print a 2-mil-wide resistor and then trim it becomes a different problem. Before considering screening and trimming, the circuit has to be converted into a series of artwork masters. A separate master will be required for conductors, for resistors (including a separate one again for each value of resistive paste) and for screened dielectrics.

The artwork is produced at either ten or twenty times the final size, depending on the degree of accuracy required and the size of the finished substrate. Then the artwork is photographically reduced down to the exact size required. There is no reason why artwork should not be produced at any size that is convenient, since it will be reduced to final size anyway. However these two are the most common reduction ratios.

The reason for working with such a large size is twofold. First, it is much easier to work with. It is easier to produce intricate artwork 20 inches square and to reduce it twenty times, than it is to work with artwork 1 inch square. The second reason is to improve the accuracy on the final photograph. As the desired pattern is photographically reduced, so is the error in drawing it in the first place. If the accuracy of artwork produced at twenty times its final size is ± 10 mils—and that is pretty sloppy—then when it is reduced the final error is only one twentieth of that or ±½ mil, assuming that no inaccuracy is introduced in the photographic process.

Preparing the Master Sheets

After the circuit is laid out, the individual master sheets are prepared. This can be done in two ways: with tape on mylar or by cutting and peeling film. In the tape method the starting point is a sheet of mylar or other clear plastic sheeting. The pattern is laid out on the sheet using pressure-sensitive opaque black or clear red tape. The tape comes in rolls of various widths and precut shapes such as holes, squares, corners, and circles.

The more common method is using cut-and-peel or cut-and-strip film. This film goes under the trade names of Rubylith* (red), Amberlith* (orange), and Stabilene.** It consists of two sheets of plastic: a clear plastic backing overlaid with a colored laminated plastic coating. The idea is to cut only through the tinted layer in the shape of the desired resistor or conductor paths and then peel away the undesired tinted portions, leaving a pattern of color on the clear backing. Figure 2-1 shows resistor and conductor patterns ready for photoreduction.

The line film used for photoreduction is not red-sensitive, so the red will photographically be the same as the black tape in the first process. Both will photograph black on the film.

The methods for cutting Rubylith are as varied as the accuracy desired. The simplest and cheapest way is with an x-acto*** knife or a razor blade.

* Rubylith and Amberlith are trademarks of the Ulano Company.
** Stabilene is a trademark of the Keuffel & Esser Company.
*** x-acto is a trademark of the X-Acto Precision Tools.

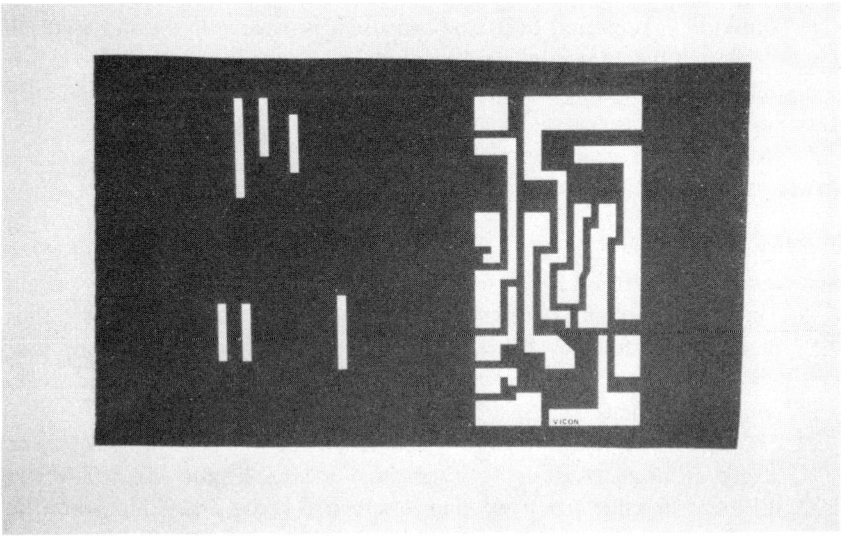

Fig. 2-1. Cut and stripped Rubylith ready for photoreduction. Resistor pattern is on the left, conductors on the right.

With a good drafting or light table, hand cutting can give accuracies of about ± 10 mils. For more accurate work a coordinatograph is desirable. This consists of a light table three or four feet or more across, with a knife blade that is very accurately controlled in the direction of the vertical and horizontal axes by a combination of gears, threaded screws, or sliding bars. The best way to produce artwork is with the coordinatograph because of the superior accuracy which can be achieved with it. Machines are available with tolerances of ± 1 mil over a cut of several feet. For even greater accuracy, numerically controlled coordinatographs are available that will automatically control the knife cut and cut depth and accurately control the spacing. Light pens are available for these so that the Rubylith cutting step is bypassed and the pattern is "cut" directly on a photographic film with a controlled beam of light.

Improved accuracy is obtained if the circuit is laid out on gridded paper. This will ensure, when the Rubylith is being cut by hand, that the cuts are all at right angles or parallel, as the case may be. The layout grid can be made to correspond to the mesh count of the screens for greater accuracy and printing control. This layout is also used to prevent sawtoothing or stairstepping when the screen stencil is made. It is important to consider the stairstepping or sawtoothing effect in screen-making before the artwork is even laid out. For the reader unfamiliar with this effect, a more complete discussion is given in Chapter 5 on screening techniques.

Rubylith is preferred over tape because it is more versatile and accurate especially when machine cutting is involved. The tape is harder to lay down accurately, must be trimmed to final accuracy with a knife, and frequently stretches or contracts.

Resistors and conductors are generally laid out at right angles in the artwork for the following reasons:

1. They can be made to conform to a layout grid for easier and more accurate layout.
2. Thick film screens have a mesh that is woven at right angles. It is desirable to line up the edges of the pattern with the mesh, thus preventing a pattern edge from running between the meshes and weakening the stencil.
3. The value of thick film resistors will vary depending on the direction of screen printing. A resistor screened along its length will be slightly different in value from one that is screened across its width. Since the possible variations in printing from batch to batch and substrate to substrate are greater across a short, wide resistor than along a long, narrow resistor, it is preferable to screen a resistor in the long direction. If the resistors are all going to be trimmed to tolerance before use, this factor is not as important as in a system that tries to hold the tolerance within the screening process and without trimming.
4. Having a resistor at an angle to the screen mesh will cause sawtoothing effects because the resistor slides cross the open areas of the screen mesh instead of lining up with the mesh at the edge of the pattern.
5. It is easier to trim resistors in only two directions. If each resistor is laid out in a different direction, it will take longer to reorient the trimming machine to take care of this.

In using either the tape or the Rubylith method, it is preferable to have more than 50% of the total area opaque (or red in the case of Rubylith), whether it is the open areas or the patterns themselves. When the artwork is photographed there is less likelihood of glare and internal reflection. One more advantage gained by using Rubylith instead of tape is that it is just as easy to strip out the open areas as it is the resistors or conductors.

It is important that the environment in which the artwork is prepared be carefully controlled. Changes in relative humidity and temperature affects the stability of the artwork. For example, an increase of 20°F in temperature increases the size of a vinyl sheet about 0.07%. This may not seem like much, but it means that a change from winter to summer in a poorly controlled room could alter a 20-inch piece of artwork by 13 mils—enough to alter the registration of the different patterns. An increase in relative humidity of 20% can change the same artwork by 6 mils. Together, humidity and temperature could have disastrous effects. A preferred environment is kept at 70° to 75°F with approximately 50% relative humidity. If these averages are not obtainable, the temperature and humidity should be kept as constant as possible. Indeed, more serious than the actual temperature and humidity (as long as it is kept within reason) are large changes in these two factors. If the temperature changes 10°F from morning to night, then artwork prepared at different times of day will be different sizes.

Storing artwork is important too. It should be stored flat in folders or between sheets of paper, preferably in a humidity- and temperature-controlled environment. The temptation is to roll up the artwork and store it in a cardboard tube or with a rubber band around it. This can have permanent effects on the stability of the pattern. Under stress the tape or red film will tend to deform or creep on the plastic backing. It can also have disasterous effects on close tolerance work.

Thick Film Pastes

Thick film resistor pastes are specified in sheet resistance values of ohms per square. This means that a paste of 1000 ohms/square, if screened into a one-inch-square resistor pattern, will have a resistance of 1000 ohms from one side to the opposite side. It will also be 1000 ohms from side to side if the square resistor is one centimeter square or 100 mils square. The size is not important as long as the resistor is square. Since this is quite a difficult concept to understand, let us study it in some detail. Consider a square pattern of 1 inch along each side and assume that the resistance is 1 ohm from side to side. This is shown graphically in Fig. 2-2(A).

Now place four of these squares together to form a bigger square 2 inches by 2 inches as in Fig. 2-2(B). This can be split into two rectangles each with two 1-ohm squares (C). Each of these can be combined to give two 2-ohm resistors in parallel (D). According to the rules for combining resistors in parallel, these two can be combined to give a 1-ohm resistor (E). The resistors could be split up and combined in the other direction if desired, with the same result. See Fig. 2-2(F) through (I). And, of course, this concept can also be extended to other sizes. A 5-inch-square resistor can be divided into five parallel branches of 5 ohms each; these can be combined to form a 1-ohm resistor again. This is the basic concept of thick film resistors. The paste values are expressed as a certain number of ohms per square. The size of the square does not matter.

To make a resistor of a fixed value, a certain number of squares are placed end to end. The actual number depends on the sheet resistance of the paste used and the resistance desired. For example, to make a resistor of 7000 ohms using a paste of 1000 ohms/square the resistor would have to be seven

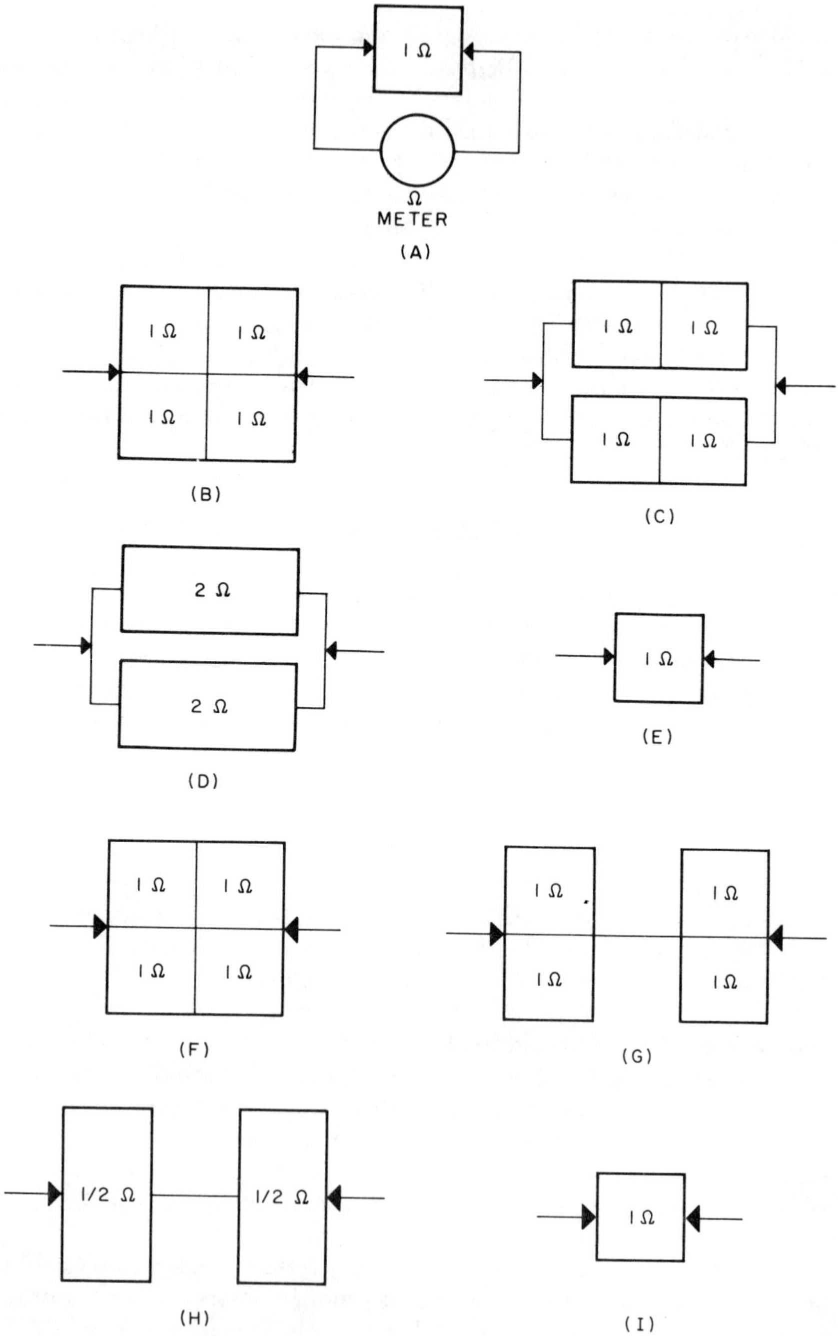

Fig. 2-2. Concept of sheet resistance and ohms/square.

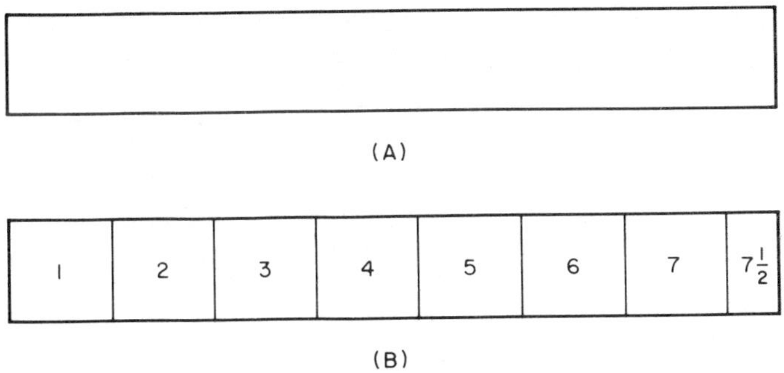

Fig. 2-3. Resistors may be divided into squares to calculate resistance. Resistor (A) divides into seven-and-one-half squares (B). If the sheet resistance were 10,000 ohms/square, the resistance would be 75,000 ohms.

squares long. This would be the same as connecting seven 1000-ohm resistors in series. Similarly, to make the resistor 2000 ohms or 9000 ohms, the number of squares would be two or nine respectively. Figure 2-3 shows how a resistor model may be divided into squares to calculate resistance.

Making thick film resistors is a matter of geometry. Starting with a paste that has a fixed sheet resistance, add squares until the resistor is the correct value. The resistor artwork pattern used will be slightly longer than the value calculated to allow for an overlap by the conductors. This gives a good electrical connection to the ends of the resistor. Sheet resistance values for paste typically come in multiples of ten, such as 100, 1000, 10,000, 100,000 and 1,000,000 ohms/square. Intermediate values are also available in many cases, and if this still does not meet the requirements, nearly any sheet resistance desired can be achieved by blending various amounts of the available stock values.

The effects of various processing conditions on resistor pastes are checked by making up test patterns. These are printed patterns of resistors in various sizes and dimensions. Such a pattern is used for checking pastes, printing conditions, printing consistency, geometry effects, firing conditions, screen-making, paste comparisons, and any other aspect of resistor-making. A typical pattern may contain resistors to give a twenty-to-one spread, from one-half square up to ten squares, in different sizes. Two test patterns are shown in Fig. 2-4.

A resistor is sometimes referred to by its aspect ratio. This refers to the ratio of length to width of the resistor; it is the same as the number of squares.

There are limits on the amount of squares that should be used to make a resistor. For reasonable results a resistor should be kept greater than one-third of a square, but less than ten squares. Outside this range of values it

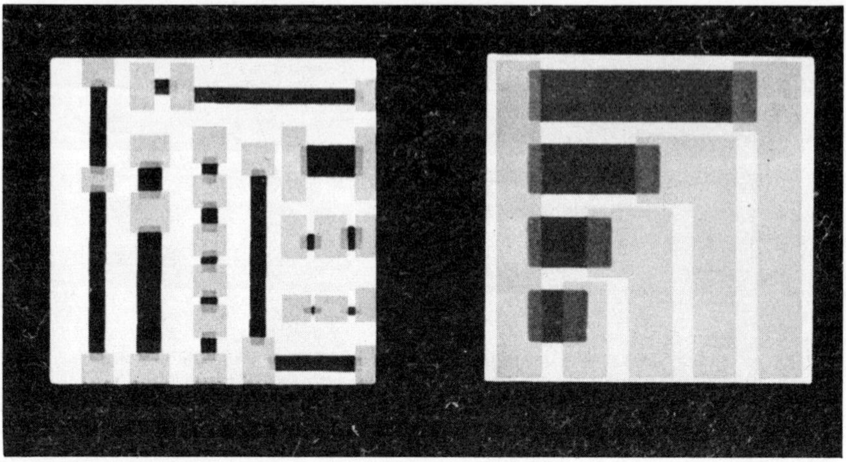

Fig. 2-4. Two resistor test patterns on 1" x 1" alumina substrates. Patterns are used to check screening and firing of various lengths and widths of resistors. Conductor is palladium-silver and does not photograph well against the alumina.

becomes difficult to accurately and repeatably screen good resistors. A better practice is to keep the limits from about one-half square to six or seven squares. These limits may have to be exceeded to meet some design requirement, but bear in mind that difficulties may be encountered.

Wattage ratings for screened resistors are difficult to specify. This depends on the material used for the substrate, the component density on the substrate, the way the substrate is mounted, and the circuit environment. General values are between 15 and 60 watts per square inch. A safe value to figure is about 25 watts per square inch. With proper care 100 or 200 watts per square inch may be attained.

A final consideration is the trimming allowance. Thick film resistors can be made to a tolerance of 10 to 20% repeatably. To reach a closer tolerance such as 1 or 2%, the resistors are trimmed to value after firing. Most of the trimming methods raise the resistor value as trimming takes place. To ensure that the resistors can be trimmed, the resistors are initially designed to be low in value. The design and layout value is usually about 70 or 80% of the desired final value. To achieve a 100-kilohm resistor at ±1%, initial layout might place it at 75 kilohms. The screened and fired value would probably fall between 60 and 90 kilohms. This can then be trimmed to 100 kilohms. If the resistor had been designed at 100 kilohms, the value after firing could be between 85 and 115 kilohms, meaning that some or all of them could be too high and not be trimmable.

Guidelines for Layout

The following guidelines should be kept in mind when designing or laying out a thick film hybrid circuit.

Minimize crossovers. Crossovers require extra screenings if a dielectric and overlay metallization is used and add the possibility of pinholing and short circuits through the dielectric. The second possibility is to use a wire bond to form a jumper across a conductor, creating two extra bonds, decreasing reliability, and increasing the danger of shorts if the wire collapses. For best protection the crossed conductor should have a dielectric coating on it in case the wire does collapse or sag. A third possibility is to use a discrete short (see Chapter 9) which adds an extra component.

Design out inductors and transformers wherever possible. Inductors for thick film circuits are available only in small values. Subminiature inductors and transformers are available, but are quite bulky and usually require soldering or welding in place.

Avoid holes in the substrate. Holes, whether for through-connections or for mechanical reasons, are expensive. Holes require a nonstandard substrate and special tooling by the substrate manufacturer. This will increase the cost and can also create long lead times due to the special tooling.

Use only one side of the substrate. Circuits have been made using both sides, but this involves more complex manufacturing and usually requires holes for through-connections, which is undesirable from the substrate manufacturing point of view.

Lay out the resistors and conductors at right angles to prevent sawtoothing and to improve the trimming ease. If possible, lay out resistors so they will all be screened either lengthwise or crosswise to further improve screening accuracy.

Capacitors take up a lot of room. Minimize capacitance and voltage ratings as much as possible. Wherever practical, design out capacitors using a technique such as direct coupling or place them outside the package.

Stay away from very high or very low resistor values. They are more difficult to process and produce accurately. If there is one very high or very low resistor in the circuit, an extra screening and drying step for that one resistor will be necessary.

During circuit design stages try to keep resistors within a reasonable range of values. If more than one or two resistor pastes have to be used, additional screening and drying steps are involved. Try to keep resistor screenings to two or a maximum of three.

Avoid long wire bonds. These can sag down onto the substrate and short out to components. They are also more likely to break if the package is subjected to sudden movement. Do not make bonds directly between two dice. A better practice is to use a bonding pad between them.

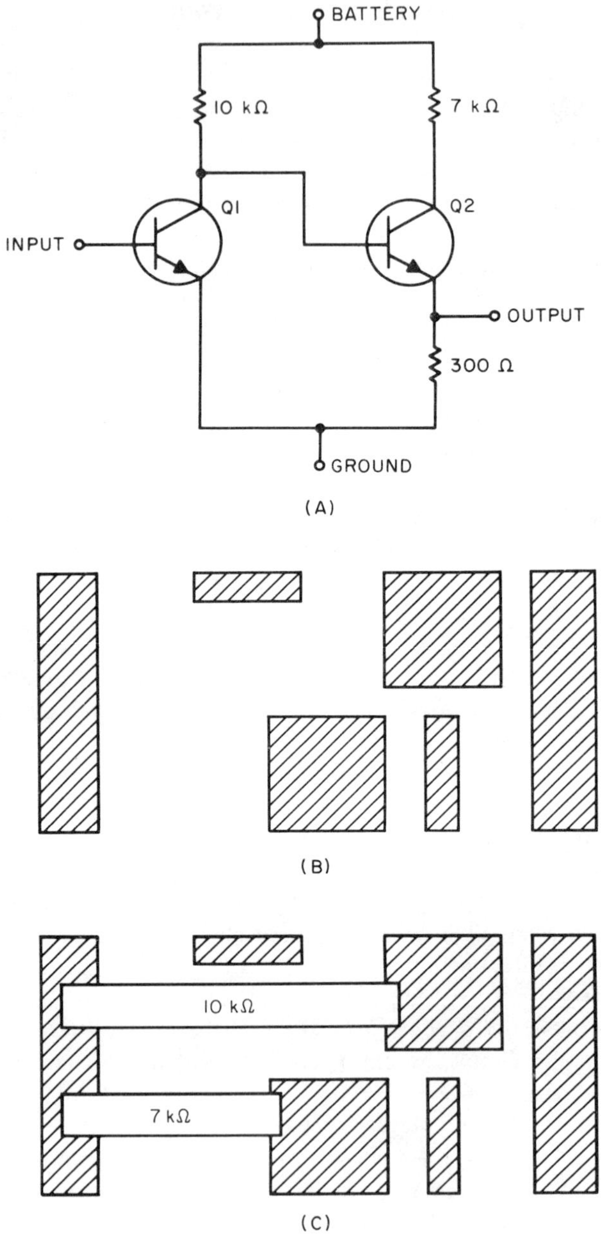

Fig. 2-5. (A) Two-stage amplifier. (B) Conductors are screened onto the substrate. (C) 1000-ohm/square paste is screened onto the substrate.

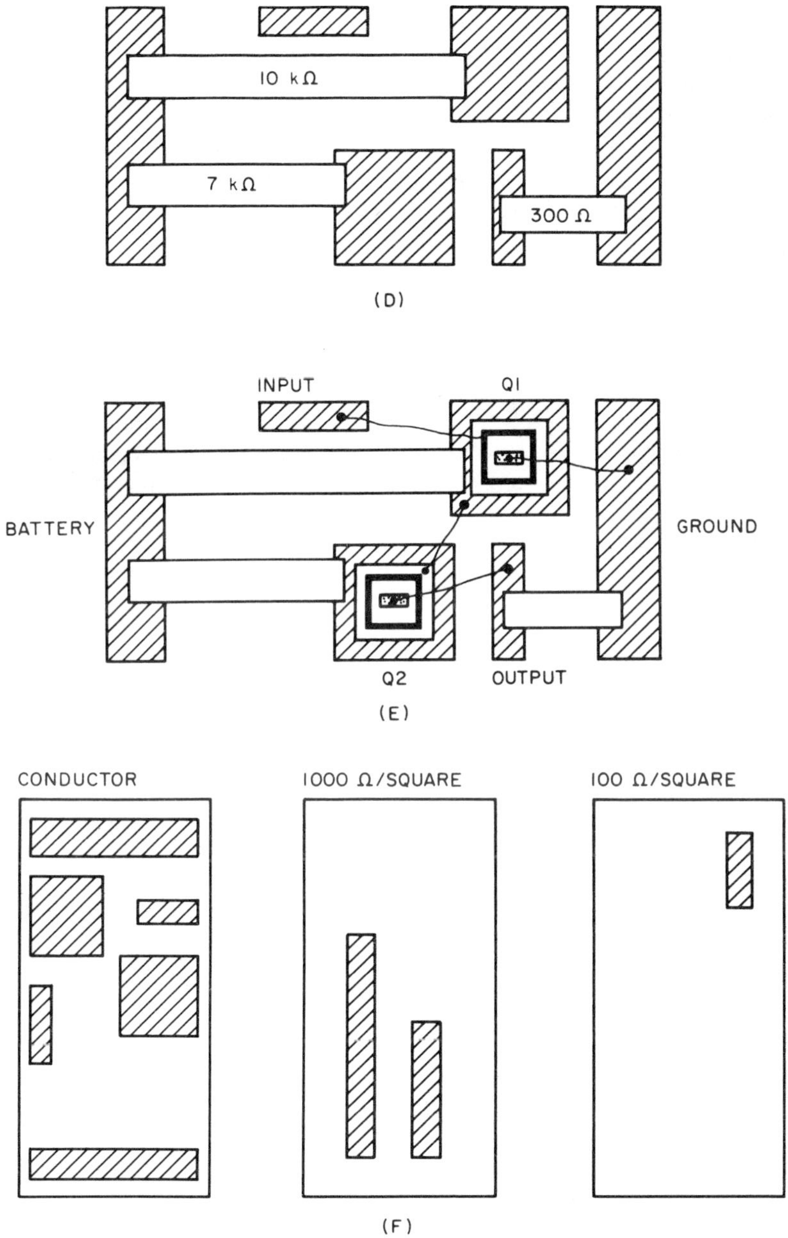

Fig. 2-5. (D) 100-ohm/square paste is screened onto the substrate. (E) Dice and wire bonds are added to complete the circuit. (F) Separate screen patterns used to make the circuit.

Care must be taken with discrete devices laid over paths as crossovers. Either the device must have its own insulation or dielectric layers must be added to ensure no electrical contact between the device and the resistor or conductor underneath.

Avoid long conductor paths. Although the sheet resistance of conductor pastes is very low, it can become enough to interfere with circuit operation if the paths are very long and thin.

The following dimensions are given as guidelines:

Conductor widths: 20 mils or greater, 5 mils minimum.
Resistor widths: 20 mils or greater, 5 mils minimum.
Conductor spacing: 15 mils, 5 mils minimum.
Resistors should overlap conductors by about one-half square.
Leave at least a 5-mil border around active device chips.
Wire bonding pads should be a minimum of 5 mils square.
Solder pads should be a minimum of 40 mils square.

A Design Example

Consider building the simple two-stage amplifier shown in Fig. 2-5 in a thick film circuit. The transistors could be in miniature form, chip form, LID package, or flip-chip or beam-lead packages. Bare dice or chips are the commonest at this time, so these will be used.

The paste choices would be 1000-ohm/square paste for the 10-kilohm and 7-kilohm resistors, and 100-ohm/square paste for the 300-ohm resistor. The circuit shown requires resistor geometries of three, seven, and ten squares. There may be maximum substrate size limitations entering into the resistor-width choice to enable the circuit to fit onto the substrate. As arbitrary values, choose a resistor width of 25 mils and a conductor width of 20 mils. This means that the resistor sizes are 25 by 75 mils, 25 by 175 mils, and 25 by 250 mils. Allowing one-half square at each end for conductor overlap these will become 25 by 100 mils, 25 by 200 mils, and 25 by 275 mils. Die bonding pads are 10 mils larger than the dice (assume 20 mils square) to allow a 5-mil border on each side.

One possible layout is given in Fig. 2-2(F), broken into conductor and resistor patterns. The three patterns are then used to make the three screens needed to screen the substrate.

3 — Photographic Processes

After the artwork has been finished at its original layout size, it has to be photographed to bring it down to its final size. This requires the use of a specialized camera for precise reduction work and a special photographic film emulsion. Figure 3-1 compares a Rubylith master to the reduced photopositive. Figure 3-2 shows reduced resistor and conductor patterns.

Basic Photography

The photographic process is quite complicated, but can be simplified to several basic steps.

A camera consists of a lightproof housing with some photographic film on one side. On the opposite side is a lens which focuses the object to be photographed onto the film, producing a clear, sharp image. A simplified view of a camera is shown in Fig. 3-3. The lens has a shutter which either shuts all the light out completely or allows it to enter the camera and strike the film for a predetermined length of time. There is a diaphragm behind the shutter (or in front depending on the particular construction) which can be opened or closed to vary the amount of light that is allowed to enter, when the shutter is open.

The focal point of a lens is the point where a beam of parallel light rays will focus into a spot. The focal length of the lens is the distance from the focal point to the center of the lens. The focal length of a lens determines, in part, how much light the lens can gather up and deliver to the film. The amount of light delivered also depends on the diameter to which the diaphragm is opened. These two characteristics are combined together to determine the f-stop for the lens-diaphragm combination. The f-stop is the ratio

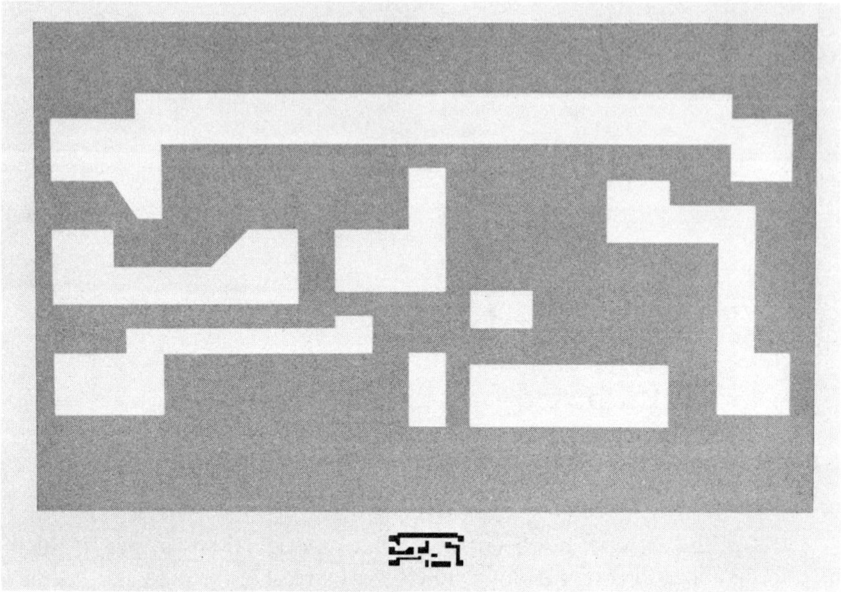

Fig. 3-1. Rubylith master conductor pattern compared to the reduced photopositive.

between the effective diameter of the lens as determined by the diaphragm and the focal length of the lens.

$$\text{f-stop} = \frac{\text{effective diameter of lens}}{\text{focal length}}$$

The diaphragm is calibrated in f-stops to correspond to standard exposures. Major f-stops are 1, 1.4, 2, 2.8, 4, 5.6, 8, 11, 16, 22, 32, 45, 64. There are two more f-stops between each of the major f-stops, called intermediate stops. It is most unlikely that a single camera will have all of these stops. A fairly typical range for a 35mm camera with a 50mm lens might be f2.8 to f22.

The other factor that determines how much light strikes the film is the shutter speed. The shutter is calibrated to open to allow light to enter, to stay open for a fixed amount of time, and then close again to shut off the light. Thus the total amount of light striking the film is determined by the shutter speed and the f-stop.

Two things to remember about f-stops are: (1) the larger the number of the f-stop, the smaller the opening of the diaphragm; (2) every time the diaphragm is moved to a smaller aperture by one major f-stop, the light

Fig. 3-2. Reduced photographic positives for a thick film circuit. Conductor pattern is on the left, resistors on the right.

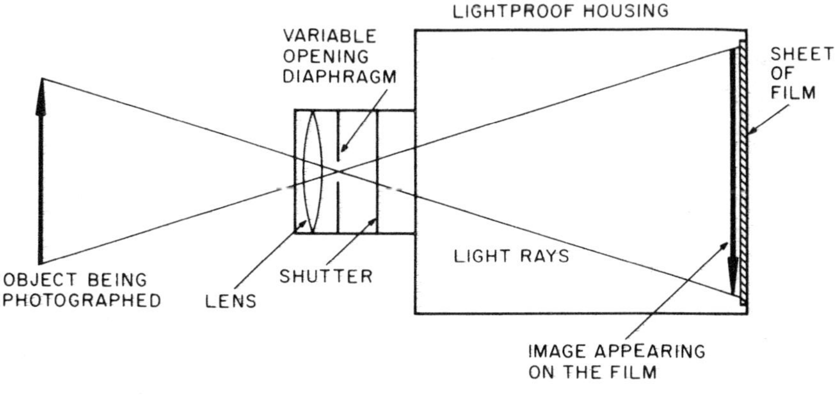

Fig. 3-3. Elements of a simple camera. Note that the image appearing on the film is upside down and laterally reversed.

admitted to the film is halved; thus to have the same amount of light striking the film, the shutter speed (length of time the shutter stays open) must be doubled. That is to say, if the photograph were being taken at f11 with a shutter speed of 1/250th of a second, and if the diaphragm were then moved to f16, the shutter speed would have to be decreased to 1/125th of a second to allow the same amount of light to reach the film. This is partly the reason why shutter speeds increase by double the speed at each setting.

Now that the light is in the camera, past the lens and shutter, it strikes the film. Camera film consists of a base sheet of plastic, such as Mylar* or Estar,** coated with an emulsion of silver bromide grains in a binder of gelatin. When light strikes these grains, a photochemical change takes place, the amount of change depending on the intensity of the light. Then, when this exposed film is placed in a special chemical reducing agent called the developer, the silver bromide is chemically reduced to form black silver grains. The amount of reduction depends on the amount of change in the silver bromide brought about by the incident light during the exposure of the film. In areas where the light was the strongest, the blackest silver will appear. In an area where there was no light to change the bromide, there will be no reduction and that area will remain white. In this manner a negative image is produced on the film.

To enable one to view and use the negative the unchanged silver bromide grains must be washed away since they too would eventually turn black if left on the film and exposed to light. To prevent this undesirable change the negative is "fixed" by washing it in a solution of sodium thiosulfate or photographer's hypo.

Although this is a somewhat crude description of the development process, it will serve as a background. In practice, the film contains more than just silver bromide grains, but also has anti-light-fog agents, stabilizers, speed modifiers, and other items. The processing chemicals contain hardening agents, wetting agents, and other ingredients to improve and modify their performance.

Photography in the Thick Film Process

With the preceding brief introduction to photography the entire process for reducing the artwork to a working photonegative can be followed.

In normal black and white photographs there are various shades of gray, called tones, ranging from white through the grays to black. These gray shadings are interpretations of the world around us as seen by the particular emulsion of the film used. The emulsion used for this type of photography is

* Mylar is a trademark of E.I. du Pont de Nemours & Co., Inc.

** Estar is a trademark of Eastman Kodak Company.

called a panchromatic emulsion. It gives a fairly natural range of tones, similar to a black and white television.

The hybrid artwork, on the other hand, consists of two distinct tones—clear open areas with red conductive and resistive patterns running around in them or vice versa depending on which is stripped off. For good results in screen-making the object is to get the maximum definition between these two tones. The emulsion used is designed to accentuate this definition. The film used is an extremely high contrast orthochromatic film—called a line film. This means that when the film is processed after exposure, any dark areas will be absolutely black and all of the clear areas will be completely clear. The film converts whatever it sees into a series of black or clear areas. There are no gray areas at all.

The emulsion can be used either on a plastic base as in normal film or on glass plates. The glass plates are considerably more expensive and are generally used in monolithic artwork reduction where extreme accuracy and stability are required. For hybrid use the plastic-based films are adequate. Estar-based films are quite stable under temperature and humidity changes and the backing is only 4 mils thick, which can be an advantage in contact printing.

The films are usually "slow" speed. This means that it takes more light or a longer exposure to achieve the same negative density as a "fast" film. This may seem like a disadvantage, but it is not. The amount of light can be controlled and since the artwork is not moving, a long exposure time can be used. The advantage here is in film resolution. One of the limitations of film (excluding camera and lens errors) is the size of the emulsion grains, the silver particles in the emulsion. If these particles are sufficiently large, then the size of the image placed on them by the lens may be smaller than the size of the grain. A single grain will try to be black and white at the same time. Because the film is high contrast film, the grain can only be one or the other. Whichever it decides to be in this case, the negative will not be a true reproduction of the image. The emulsion grains are the individual parts that make up the picture and, as they cannot resolve anything smaller than themselves, the image is lost or distorted. Film speed goes hand in hand with grain size. The higher the speed of the film, the larger are the clumps of silver particles and the more the individual grains will be noticed in the final negative. In hybrid photography this would distort the edges and corners of the pattern, thus destroying the fine detail required. Therefore high resolution line films are generally slow speed and very fine grain films.

The film is placed in the camera either in a cut film holder or glass carrier, or is held with a vacuum plate, which ensures that the film is held absolutely flat and eliminates the tendency to curl that can occur when using cut film holders. Curling at the film edges can produce slight inaccuracies in the negative.

Equipment and Environment

The cameras used for thick film reduction work are the larger sizes of sheet film camera, usually 5″ x 7″ or 8″ x 10″, mounted on rigid rails to provide accuracy of alignment. There are several reasons for choosing these sizes over smaller sizes such as 35mm or 2¼ x 2¼ roll film cameras:

1. Single sheets of film are easier to handle and are more efficient since a whole roll of film does not have to be used to process one shot.
2. The larger size is needed for making intermediate reductions. If a step-and-repeat is to be done at five times reduction from a 20-inch piece of artwork, then the extra film space is needed.
3. Glass plates can be used with the larger cameras.
4. Sheet film comes cut in flat sheets that do not have the tendency to curl that roll film does. This curl can be a source of inaccuracy in shooting or in contact printing. The small amount of curl encountered in sheet film can further be controlled by the use of vacuum hold-downs that keep the film absolutely flat.
5. Most of the time the entire sheet of film is not used, only the central portion. The image will come through the center part of the lens. This will help eliminate any distortion effects which become more pronounced at the edges of the lens.

For shooting a negative, the artwork is placed on a flat, lighted area called the copyboard. The artwork is held to the copyboard by either enclosing it under a sheet of optically flat glass or by holding it flat against the copyboard by a vacuum. The vacuum method is preferred since it eliminates two glass surfaces between the lens and the artwork.

The copyboard itself can be lighted by either of two methods. The first is the reflected light method, as shown in Fig. 3-4(A). One or more lights are placed at an angle to the copyboard on each side of the artwork. They are shielded from the lens to prevent undesired direct flare into the lens. The light sources used can be photofloods, though for more critical applications, arc lamps or quartz-iodine lamps should be used. Photofloods change their light output characteristics as they age and only have a working life of 4 to 6 hours.

For best results it is important to have even lighting across the copyboard. The old rule of thumb is to have the lights at a 45° angle to the copyboard and evenly spaced at each side. This is not the optimum case as the center of the artwork will receive more light than the sides. A better scheme is to mount the lights at an angle of about 30° to 35° to the copyboard. This results in more even lighting. Even lighting becomes more of a problem with large pieces of artwork. In these cases the edges of the artwork

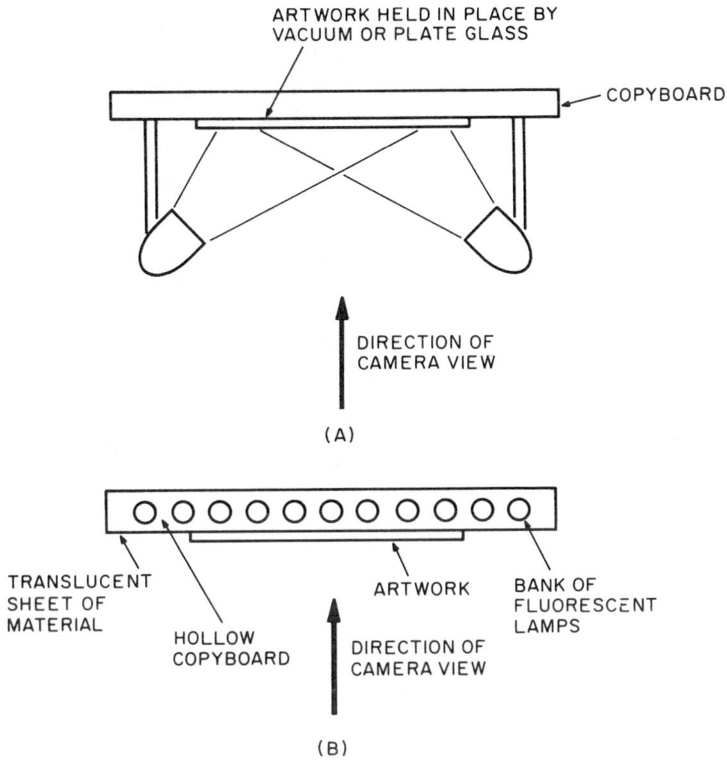

Fig. 3-4. Copyboard lighting techniques with front lighting (A) and back lighting (B).

are actually farther away from the camera lens than the center of the artwork. This will require further manipulation of the lamps to provide even lighting. In practice, copyboard lighting is not as much of a problem as it sounds. With the size of artwork and film dealt with in thick film work, uneven lighting effects are usually negligible.

The second type of lighting is called back lighting. This consists of placing lamps behind a translucent copyboard and shining a diffused light through the artwork as shown in Fig. 3-4(B). This is an excellent technique because of the opaque-clear characteristics of the artwork. Back lighting produces better results than front lighting. Front lighting will give a contrast ratio (relative contrast between light and dark areas) of about 40 to 1, whereas back lighting can give a contrast ratio of 1000 to 1. This obviously increases the photographed definition between black and clear areas. Light sources for back lighting can be photofloods, arc lamps, or the favorites—a bank of fluorescent tubes. Fluorescent tubes run cool, are cheap, and provide a very good, white light.

Focusing of the image on the ground glass is done with the lens wide open (diaphragm at its widest opening). Focusing can be aided by using a low-power microscope or magnifier to check the image. When the negative is actually shot, the lens is "stopped down" or set to a smaller diaphragm opening. Using a smaller than wide-open aperture or diaphragm opening will give an increased depth of focus. The depth of focus refers to the range of distance over which objects are in focus. Stopping the diaphragm down will increase the depth of field, compensating for any small focusing errors or inaccuracies in the alignment of the copyboard, lens, and film planes. If the lens is stopped down too far, the reverse is achieved. The definition becomes poorer again. The best definition is obtained at two or three f-stops less than the maximum f-stop, usually around f16 or f22.

Exposure time and f-stop value are best obtained by the use of a reliable exposure meter once the desired lighting conditions are established. If a meter is not available, a series of shots at different speeds and f-stops can be made until the desired negative density is achieved. Underexposure has occurred when the black areas are still gray and are not completely black and opaque. The film is overexposed when the clear areas start to turn black, lose their definition, and blur around the edges.

On a camera system set up for precise work, the reduction ratio is measured on a calibrated track. Figure 3-5 shows a simplified view of a reduction camera on a track. To achieve extreme accuracy, the adjustments are made with micrometer screws and a low-power microscope. If the reduction is being done on a homemade system without a precise track, then the method is simply to measure the distance between calibrated registration marks on the artwork on the image in the ground glass focusing screen. This

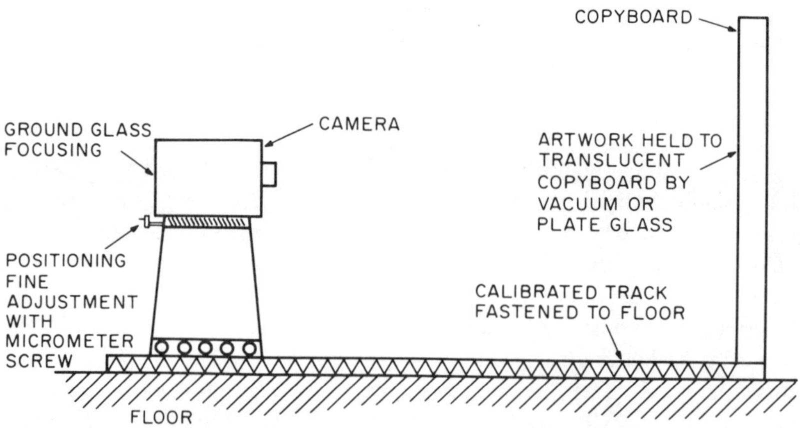

Fig. 3-5. A simplified view of a camera set up for artwork reduction.

will have to be done with care and may take a couple of test shots due to the inherent inaccuracy in measurement caused by the thickness of the ground glass used for viewing and focusing.

There are several important points to remember in camera system specification.

1. The lens to use is a high-resolution lens of good quality. One of the more annoying defects in a lens produces optical distortion of the straight lines being shot. This is either barrelling or pincushioning, as shown in Figs. 3-6(B) and (C). These are the results of uneven magnification across the lens. This is very bad in hybrid work, since the accuracy of the circuit depends in part on the straightness of the lines. This will not occur in a high-quality lens designed for line photography. The distortion can be minimized by using a large camera and only a small image in the center of the film. This means that the edges of the lens are not being used, so this type of distortion, if present, is not as pronounced.
2. The copyboard must be perfectly flat, so the artwork is all in one plane. Misfocusing and wavy lines occur if it is not.

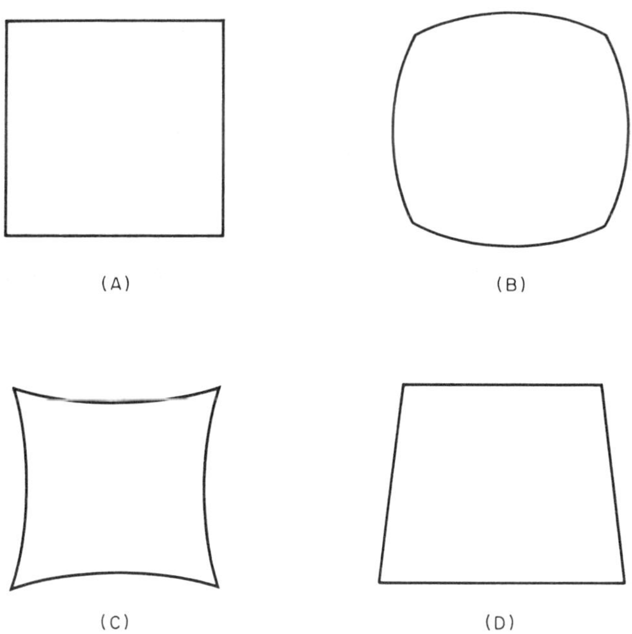

Fig. 3-6. Common photographic distortions. (A) True image. (B) Barrel distortion. (C) Pincushion distortion. (D) Keystone distortion.

3. The film and copyboard planes must be exactly vertical and parallel. If they are not, the distance from the corners of the artwork to the corners of the image will vary, leading to a difference in focusing distance in various places and uneven reduction of the artwork. A most common distortion is keystoning, shown in Fig. 3-6(D). This occurs when the camera back and copyboard are not exactly parallel.
4. The lens axis must be exactly perpendicular to the copyboard-film axis. Optical distortion results if it is not. To ensure that the camera and copyboard are aligned exactly, the camera is usually mounted on a rigid track or on sliding rails. These are mounted securely to the floor. The copyboard is either fastened directly to the end of the track or is bolted to the floor after initial alignment. Various reduction ratios are achieved by moving the camera up and down the track to vary the distance between the camera and the copyboard. An alternate scheme is to mount the camera securely in one place and use lenses of differing focal lengths for different reduction ratios. This last scheme results in a few fixed ratios, whereas the sliding rail or track can provide any ratio desired.
5. The camera room must be dust- and fume-free. Dirt and chemical fumes are very fast ways to ruin fine optical equipment. Dust on the artwork or film or camera lens will create pinholes in the final photograph. A common problem, especially in low-humidity environments, is static electricity in the artwork, causing dust to cling to the plastic. This can be eliminated by wiping the artwork with chemically treated cloths similar to phonograph record static eliminators, or by using special antistatic sprays.
6. Temperature and humidity should be controlled. A temperature of 70°-75°F with a relative humidity of approximately 50% is good. Temperature variations, in particular, can create inaccuracies. Just the expansion of the camera tracks with a temperature rise can alter the reduction ratio.
7. Vibration must be controlled. This is one factor often overlooked. Heavy machinery operating in a nearby location can create so much vibration in the camera that blurred edges on the final negative can result.

It must be kept in mind that all the above precautions may not be necessary. It depends on the accuracy desired. Perfectly acceptable thick film negatives can be shot using two desk lamps and a press camera. If this is adequate, fine. However, as the degree of sophistication increases, as finer and finer lines are required and precise alignment becomes a problem, all these factors must be considered.

One more facet of picture-taking should be considered. That is the step-and-repeat process. Step-and-repeat photography is the production of a series of images, precisely spaced on one negative. The reason for this is to permit multiple screen printing on one substrate. Several circuits are screened at one time on one substrate. Then, after the final firing, the substrate is scribed and broken into separate circuits. This effectively increases production rates over the one-at-a-time process.

The step-and-repeat process uses either a single- or multiple-lens technique. In the single-lens technique, the film is accurately moved in the direction of the x and y axes with a precise mechanical system, and the film is exposed at each desired location. In the multiple-lens technique, several lenses are mounted on the lens board; each one is focused on a different area of the film. This repeat process is made all in one shot. Often a first reduction of the artwork is made, then this is stepped-and-repeated. A second reduction is made of the stepped-and-repeated pattern to reduce any step-and-repeat error, and produce the final pattern. If the reduction ratios are small, then only a single reduction would be made.

The final photographic process to be considered is contact printing. When the first shot is taken of the artwork, a negative results. That is to say, the clear areas on the artwork will be black on the film, and the red or taped areas on the artwork will become clear on the negative. The reverse of this, where the red and clear areas are the same as on the artwork, is called a positive. In screen-stencil-making for the thick film printing process, a positive is used. A method of obtaining a positive from a negative and vice versa is required. It is very easily done. The negative is placed in contact with an unexposed piece of film and held there in a glass frame or with a vacuum system. The film is exposed to light through the negative and then developed as usual. This will provide a reverse image. In this way a master negative can be kept in a safe place, and as many working positives as are needed can be produced.

For very precise work, the negative and unexposed film should be placed emulsion-to-emulsion, as diagrammed in Fig. 3-7(A). If they are placed emulsion-to-backing, as diagrammed in Fig. 3-7(B), the light coming through the negative to expose the film will diffuse on its way through the backing and thereby degrade the sharpness of the lines. The amount of diffusion, naturally, will depend on the thickness of the backing of the film being used. Contact printing produces a lateral reversal or mirror image as well as an image reversal. This brings up one advantage of using Estar-based films, where the backing is only 4 mils thick. If lateral reversal is not desired and the application is not critical, then the negative and the unexposed film can be placed emulsion-to-backing, and the amount of diffusion through the backing of the film will be negligible.

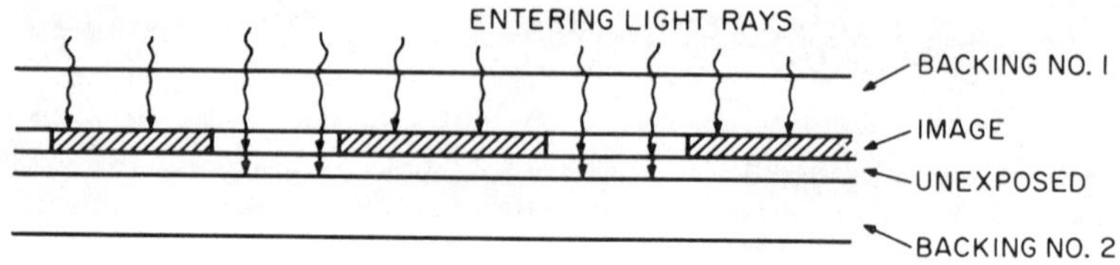

(A)

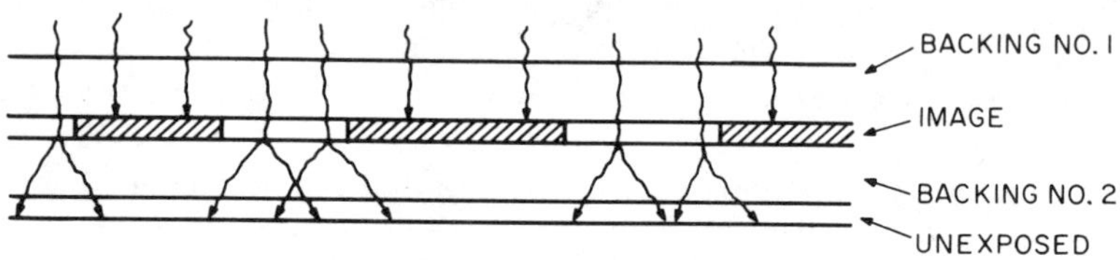

(B)

Fig. 3-7. (A) Emulsion-to-emulsion. The light enters through the backing of the exposed film. It is stopped by the exposed image, but can pass through the clear areas. Since the films are emulsion-to-emulsion, the light passes directly into the unexposed emulsion layer and exposes it. Because of the close contact of the emulsion layers, the pattern stays well defined. (B) Backing-to-emulsion. The light enters through the backing of the exposed film. Again, the exposed image stops some of the light and the clear areas pass the rest. But this time, the light passing through the clear areas does not go directly into the unexposed emulsion. It has to pass through the backing of the unexposed film first. As it does so, the light tends to become diffused and spreads out over the emulsion when it gets there. Thus, the pattern projected onto the unexposed emulsion is not a true reproduction of the exposed image. The thicker the backing of the film, the worse are these scattering effects.

Film Development

After being exposed, the film is ready to be developed. Unload the film from the camera or film holder in total darkness or use the safelight specified by the film manufacturer. Usually a dark red safelight can be used.

1. Develop. Use the recommended developer for the time at the temperature recommended for the particular film being used. The time-temperature relationship for developing is very important. The higher the temperature of the developer, the faster the development takes place, so the development time has to be shortened accordingly. Recommended temperature is usually 68°F. The film should be agitated in the developer according to the manufacturer's instructions. Agitation brings fresh developer to the film surface and accordingly increases the

rate of development. Too much agitation will increase the grain structure and give uneven development. No agitation will cause streaking and sometimes increased grain. Increased grain will also result if the film is left in the developer longer than the recommended time. Leaving the film in too long will also cause the clear areas to "burn in" or start to turn black even though they are not supposed to. It is very important to follow the film and developer manufacturers' directions. Development takes place either in darkness or under a safelight.

2. Rinse or stop bath. Stop bath is a weak solution of acetic acid, that stops development instantly when the film is immersed. It also helps to cut down on pinholing in the film. It is not essential. If a stop bath is not available, a vigorous 30-second rinse in running water to remove the developer and to stop development is adequate. This step also takes place in darkness or under a safelight.
3. Fixer. Next immerse the film in the fixing solution for the recommended time or for twice the time it takes for the film to clear. Room lights can be turned on about halfway through this step, and the rest of fixing and washing take place with normal room light. Agitate gently while the film is in the fixer.
4. Wash in running water for the recommended time.
5. Dry in a dust-free atmosphere.

It is important to follow the manufacturer's recommendation for temperature control. As well as changing development time, it too can affect the grain. The temperatures of all the solutions and the wash water should be kept close to the same. If the film is plunged from a solution at one temperature into a rinse bath at another, a reticulation effect can take place that leads to the appearance of increased grain.

Now that the circuit is laid out and reduced to a working negative, the actual hardware for the thick film circuit can be discussed.

4 — Substrates

The basic building block for a hybrid circuit is a flat, close-tolerance wafer of ceramic material. It serves as a foundation on which the circuit is built. Figure 4-1 shows several different types of thick film substrate.

Materials other than ceramic are sometimes used, such as an ordinary copper-clad printed circuit board for use with some resin-based pastes, but because of the requirements for the base material and the peculiarities of the hybrid process, ceramics work best. Ceramic materials of several types have widespread usage.

The requirements for a substrate material are:

1. Good thermal conductivity to conduct heat away from possible hot spots in the circuit, such as power transistor chips.
2. Smooth, flat surface for precise, repeatable screen printing.
3. Ability to withstand the high temperatures used in resistor- and conductor-firing.
4. A good electrical insulator, so that leakage paths between conductors do not occur.
5. Chemical inertia, because of the various cleaning agents and process chemicals that it will be in contact with during circuit manufacture. It must be nonabsorbent for the same reason.
6. Low thermal coefficient of expansion and coefficient compatible with materials to be screened or mounted on the surface.

Fortunately, these characteristics are found in ceramic material such as alumina or beryllia.

Ceramic techniques are very sophisticated and the substrates are made to very close tolerances. A wide range of geometries is possible by working with the vendor. The processing of ceramic takes place in the temperature

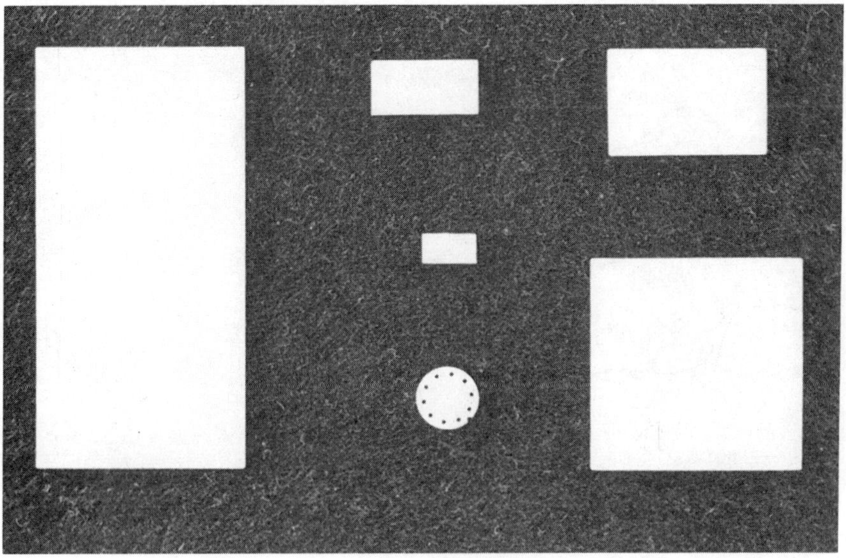

Fig. 4-1. Miscellaneous alumina substrates ranging from 1/8″ x 1/4″ to 2″ x 1″. Round substrate is for a TO-5 can.

areas of 1400°-1900°C. Resistor and conductor firing takes place from around 700°-1000°C. This means that fired ceramic will not be affected by the lower temperatures of thick film processing, thus eliminating the possibility of chemical interaction between substrates and pastes during firing.

Types of Ceramics

Ceramics come in three physical types: single-crystal form, polycrystalline form, and noncrystalline or glass form. In the single-crystal form, as the name suggests, the entire piece of ceramic is formed as a single crystal. The arrangement of the internal atoms is in a regular, repeated lattice structure. Single-crystal ceramics have very good surface properties, but they can cost ten or twenty times as much as the polycrystalline form. Aluminum oxide occurs in the single-crystal form as sapphire. This is often used as a thin film substrate because the smooth surface is excellent for film deposition. For thick film work, single-crystal ceramic is not necessary and would cost too much to be feasible. The smoothness of the single-crystal form is also actually a disadvantage. Better adhesion of pastes will occur with a slightly roughened surface.

The opposite of this, the glass form of ceramic material, does not have any crystalline structure. A glass is a rigid, fused material that does not go

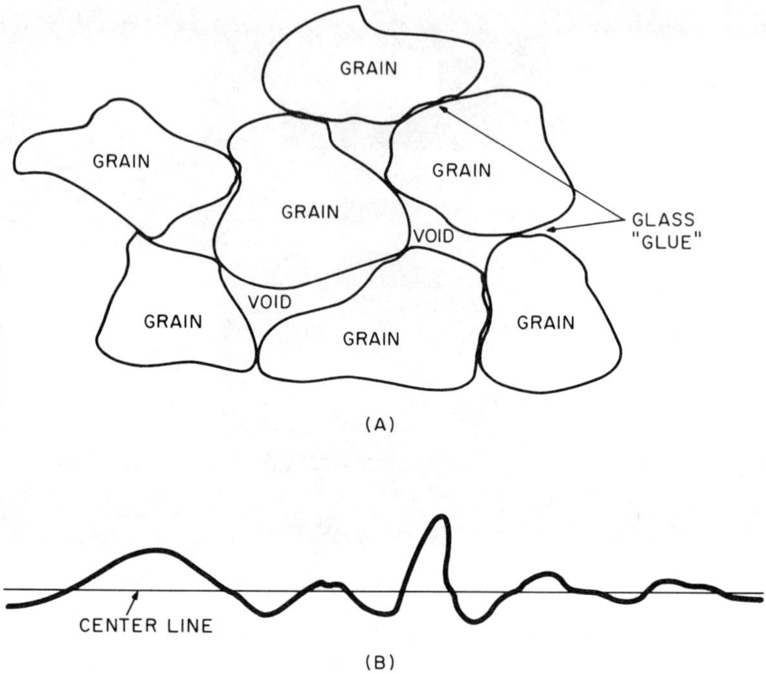

Fig. 4-2. (A) An exaggerated view of ceramic microstructure. (B) Cross section of ceramic surface.

through the crystallizing phase when cooling from the liquid phase to the solid phase. The glazes on a glazed substrate or a resistor overglaze are a form of low-melting-temperature glass that is screened, dipped, or sprayed onto the ceramic and then fused to the surface with a low-temperature firing cycle.

Polycrystalline ceramics, such as the common aluminum oxide or beryllium oxide, are the most common substrate materials. The polycrystalline form of ceramic is a combination of the single-crystal form and the glass phase. It consists of small grains of ceramic (usually single-crystal, though not necessarily) that are "glued" together by glass. These grains do not compact together to form a single mass, but have gaps between them called pores or voids. The entire surface of grains and voids is called the microstructure. An exaggerated view of the microstructure is diagrammed in Fig. 4-2(A). It is obvious that for a strong, smooth surface, the grains should be very small and close together, thus keeping the voids very small.

The properties of ceramic substrate material depend strongly on the ratios of glass, grains, and voids. The larger the voids are, the weaker the mechanical strength of the substrate will be. As the relative amount of glass in the mixture goes down, most of the substrate properties will improve. One

Fig. 4-3. A comparison of substrate materials. On the left is single-crystal sapphire and on the right is polycrystalline alumina. Both substrates are 1″ x 1″.

such property is thermal conductivity. Glass is a poor thermal conductor. As the glass content and void area decrease, the ability of the substrate to conduct heat will increase.

Polycrystalline ceramic is a conglomerate of chemicals. It contains more than just grains and glass. To further modify the properties, a controlled ratio of impurities is added. An alumina substrate is composed of aluminum oxide, Al_2O_3, ranging in purity from 85 to 99.5%, and the balance of the material is made up of impurities. Common impurities are calcium oxide, magnesium oxide, silicon oxides, and complex substances such as talc. Talc lowers the melting point of the material. Magnesium is sometimes added to suppress grain growth and increase strength. Figure 4-3 compares alumina and sapphire substrates.

Methods of Manufacturing Substrates

There are two methods commonly used for manufacturing of substrates: dry pressing and slip casting or tape casting. Both methods involve quite complicated design procedures and very close tolerances. Ceramic materials, when dried and fired, will shrink a certain amount depending on the composition and processing procedures. This shrink factor can be calculated and controlled, but the parts have to be designed oversize and the processing conditions carefully controlled to ensure even and repeatable shrinkage from substrate to substrate.

Dry pressing is the process of mixing ceramic grains with a suitable binder and squeezing the material together in a die at high pressures. Hydraulic and mechanical presses are used to deliver several thousand pounds of pressure per square inch. The material will compact and bond together. The die manufacture is quite complicated in itself. The die must withstand very high pressures of up to around 20,000 pounds per square inch. It must be an even harder material than the ceramic so that the die surfaces will not be abraded away. The press must be designed to apply equal pressure across the top of the fabricated part and account for the differing pressure effects at the sides and other die surfaces. If the pressure is not even, stresses can be set up in the substrate and cracking will occur during the firing stage.

The preliminary processing of the grains themselves is quite complicated. First, large grains act as the basic building blocks, then small grains fill in the voids, and finally a powder is mixed in with the binder to cement the whole structure together. The preparation of materials has to be extremely carefully controlled, because the size and properties of the materials affect the final microstructure, which in turn controls the physical properties of the substrate. Very close tolerance milling and grinding of the grains is involved.

Dry pressing is generally used for thicknesses greater than 50 mils. The pressing is followed by drying and then the firing or sintering step. One type of pressing, called hot pressing, combines the pressing and the sintering steps by heating the molds to between 1000° and 2000°C during the pressing. In general, the temperatures used are higher than in sintering, and special molds have to be used. Graphite is a commonly used mold material, but it is relatively soft, leading to increased mold wear and the resulting replacement expenses. This technique is more expensive than the others, and is generally not used for substrate manufacture.

Tape casting is relatively cheap and easy; it is the common method of producing substrates. The tape method involves making a ceramic paste or slurry, using an organic resin binder and appropriate solvents. This is deposited in a thin even layer on a transfer medium such as plastic or glass plates. Continuous plastic belts can be used in a mechanized process. Then the solvent, which may only be water, is carefully dried off, and the green ceramic tape is peeled from the transfer medium. The term "green" is applied to ceramic materials in the fabricated, but unfired, state.

After drying, the green tape is flexible and lends itself well to additional mechanical processing. It can easily handle any die-punching operation. A hole or hole patterns can be punched out. Any desired size or shape can be easily stamped out. As with pressed parts, substrate shrinkage during the firing stage has to be taken into account when considering dimensions. Tape casting is suitable for substrate thicknesses up to about 50 mils.

Following preliminary formation by either method, the green ceramic goes through a drying process. Tape cast parts have to evaporate out the final traces of the solvent. The pressed parts, even though they do not directly involve solvents, usually have some absorbed moisture or mold lubricants that have to be removed. Drying is carried out between 400° and 500°C, fast enough to keep the process moving, but not so rapidly that the material cracks or warps. This is one of the steps where the ceramic shrinks, due to further compaction and the loss of solvents, and must be very carefully controlled.

The final step is sintering or firing, where the green ceramic is fired at a high temperature to sinter or fuse the ceramic particles together. Either a batch kiln or a continuous belt kiln can be used. The actual temperature used depends on the purity of the material and the final properties desired. The more impurities that are present, the lower the melting point of the material. The object of the firing step is to produce a fused microstructure, suitable for the end purpose of the ceramic. Thus a delicate time-temperature relationship, similar to that seen in screened resistor firing, is involved. If the sintering temperature is too low, the ceramic does not fuse suitably. If it is too high, too much of the glass phase is produced and the substrate properties are altered. If the firing time is too short, not enough sintering occurs and if it is too long, the grain size increases too much, again altering the substrate properties. The temperature and time at peak firing have to be regulated to produce a solid, compact material with small grain. Of course, the grain size can be partly controlled by using very fine-grained material to begin with when the green ceramic is fabricated. Firing takes place from 1400°-1900°C, depending on the desired results and the impurity content of the material.

Contamination

Cleanliness is very important throughout all stages of circuit work and begins with the correct preparation of the substrate. Substrates, as received from the manufacturer, are not usually suitable for immediate use. The handling they have received since their manufacture may have covered them with dust, lint, and skin oils. Added to this are particles of talc, any oils or mold release from the processing equipment, and other contaminants that no one would suspect. These must be removed. If they are not removed, poor adhesion of the paste to the substrate will result. Adhesion strength can be increased by a factor of two or more with good cleaning. One of the best methods of cleaning is ultrasonic agitation with a quality grade of Freon* or other degreasing agent. Ultrasonics are necessary to force the dust and talc out of the micropores in the surface of the ceramic. Some cleaning agents, such as acetone, leave a film on the surface that has to be washed off by another agent, making a two-step process. Some manufacturers have tried

* Freon is a trademark of E. I. du Pont de Nemours & Co., Inc.

soap solutions, such as common detergents or dishwashing liquids. With ultrasonic agitation these are satisfactory also. Adequate steps must be taken to entirely rinse the soap from the substrates. Residues of soap or other cleaning agents can cause chemical interactions during later process steps such as the firing stage. Soap, especially, when fired, can produce chemically reducing hydrocarbons that will interact with palladium-silver systems. After cleaning, the substrates should be dried in a dust-free atmosphere using warm air or lint-free towels.

If possible, the substrates should be run through a firing cycle or a batch oven before use. This will burn off any contaminants left from cleaning that might react with the printed films. A temperature of about 800° to 1000°C can be used for this. In a batch oven, about a 20-minute soak is suitable. The substrates can also be run through the firing furnace. After the cleaning process, the substrates should be stored in a dust-free atmosphere, preferably in a sealed container, and only handled with finger cots, gloves, or tweezers.

Ungloved hands constitute a particularly bad source of contamination. Skin oils and salts are passed onto the substrate. The oils can be removed by a suitable degreasing agent. However, the salts are not removed by many of the conventional degreasing solvents. If this type of contamination is suspected, a two-step cleaning process may be necessary. A water-based detergent or detergent additive is used to remove the contamination, followed by another type of solvent to ensure that the first solvent is completely removed.

Substrate Specification

The following must be considered when specifying substrates: material, size and thickness, location of holes, if any, surface finish, glaze, and camber.

Material. The commonly used materials for thick film substrates are alumina, beryllia, and steatite. Barium titanate has a specialized use.

The industry standard for most thick film work is 94 to 96% alumina (Al_2O_3). Also available is 99.5% alumina. It is used for very precise thick film work. Surface smoothness is two or three times better in 99.5% alumina than in 96% alumina. Thermal conductivity is slightly better in the 99.5% alumina, because of the lower impurities. The 99.5% is more commonly used for thin film circuits, as 96% alumina is quite adequate for most thick film work. Consideration of cost enters into this, since 96% alumina is cheaper.

When heat dissipation becomes a problem, substrates can be made of beryllium oxide (BeO). Beryllia is considerably more expensive than alumina, but the high heat dissipation capabilities make it more desirable. Thermal conductivity of both alumina and beryllia depend on the actual process used to produce the fired ceramic. The presence of pores and the percentage of glass and impurities will alter the characteristics of the material, so it is difficult to quote specific values. Specifications should be obtained from the manufacturer for the actual substrate material being used. However, in general, thermal conductivity for beryllia is about eight to ten times better than for alumina. Beryllia has the disadvantage that the beryllia dust created by any machining is very toxic. If beryllia substrates are to be cut or drilled, or even an abrasive resistor trimmer used on resistors on a beryllia substrate, then special precautions have to be taken to prevent operators from coming into contact with or breathing any of the dust. The effects are similar to silicosis, a lung disease contracted from inhaling coal and rock dust over long periods of time.

Steatite ($MgO \cdot SiO_2$) is a very cheap substrate material. It was used as a substrate for the first thick film circuits. It is easier to work with than alumina and uses lower processing temperatures. However the physical properties are inferior to alumina and so it is generally used only for very low cost, high-volume, noncritical applications. Its thermal conductivity is especially poor.

Barium titanate ($BaTiO_3$), while not strictly a substrate material, is a ceramic material widely used in hybrid circuits. Its unique advantage is an extremely high dielectric constant. The relative dielectric constant can be up to 5000, compared to 9.2 for 96% alumina. The actual dielectric constant, like other properties of the ceramics, will depend on physical characteristics, such as the microstructure, and the manufacturing process. This high dielectric constant is used to advantage in making capacitors for hybrid circuits. Chip capacitors made of barium titanate layers are very common, and will be discussed further in Chapter 9 in the section on capacitors.

Size and Thickness. Substrates are available in a wide variety of sizes and shapes with stock tooling and infinite variations can be made to special order. It is preferable to stick to standard sizes, if possible. For nonstandard sizes, the vendor will usually make a one-time tooling charge, which may or may not be depreciated over the order, depending on the vendor and the quantity involved. Nonstandard sizes usually involve longer delivery time, since the special tooling has to be made up.

Nonstandard sizes can be made up two ways. Either they can be incorporated into green ceramic tooling and punched out to the desired size or the substrates can be machined after firing. Punching the green ceramic is preferable where large quantities are involved since it is easy and cheap. Any machine work done after firing is expensive because of the hardness of the fired ceramic. Post-fired machining can be done however, and for small quantities this would be quicker and cheaper. The ceramic can either be cut with a diamond saw or with diamond dust in a cutting slurry. For straight line cutting, diamond scribing is a popular method. The principle is exactly the same as cutting glass. A diamond point is run across the face of the ceramic to create a groove that weakens the ceramic in the desired line. Then the substrate is carefully snapped apart along this scribe line. Mechanized scribes and fracturing tools are available to produce a high-quality break. Laser scribing is becoming more popular. The laser drills a series of tiny holes in a straight line, similar to postage stamp perforations. Then the substrate is snapped apart along this line. This is a fairly quick and cheap method, after the initial equipment investment. If small quantities of odd-sized substrates are to be used, a small diamond scriber can be a good investment. This allows quick turnaround time and cheap cutting of standard-sized substrates. Diamond scribes usually come with four scribing points per tool and can be resharpened quite cheaply. Scribe life of around 100 inches per point can be expected.

Pre-scribed, but unbroken, substrates can create an effective speed-up of processing time that can be used to advantage. The circuit can be designed for processing in multiples and then broken apart after screening and firing. For example, if the final substrate size is ½″ x ½″, it may be more economical to print and fire the circuit using four identical patterns on a 1″ x 1″ substrate, and then split it into four pieces. The substrate can be bought from the substrate vendor already scribed (or this can be done in-plant), and then it can easily be snapped apart to give the four separate circuits. This technique will effectively multiply this particular processing output by four.

Ceramic thicknesses are available from about 10 mils to 100 mils. The most common thickness for thick film work is 25 mils.

Holes. Holes can be made by the vendor either in the tooling or, in small quantities, by diamond or ultrasonic drilling. For post-fired holes, the same problem exists as for post-fired sizing: fired ceramic is a very hard substance and, as a result, is very expensive to machine. Diamond and ultrasonic drills are costly to buy and use. The results are not as good as if the holes had been made in the green ceramic before firing. Laser drilling is being used, but even though the holes are cheap by this method, the initial investment is quite high and processing time longer than if punched green ceramic techniques are used. Holes are an expensive proposition as far as substrate manufacture goes; therefore it is best to eliminate them, if possible, in the original design stages.

Surface Finish. The relative roughness of the substrate is of great importance. It is desirable to have a flat, smooth surface for ease and uniformity of printing and for the performance of other operations. At the same time, the surface must have a sufficient degree of roughness to give the film of paste good adhesion strength. If the surface is absolutely smooth, the fired pastes will not adhere as well.

Because of the ceramic grain structure, the surface of the substrate is quite uneven. The surface consists of an aggregate of grains of ceramic making high peaks, and voids forming deep valleys. A cross section would show a very irregular surface. The variation between peak and valley bottom can be 80 to 100 microinches. An exaggerated view is shown in Fig. 4-2(B).

Manufacturers' specifications refer to CLA (center line average). This is a hypothetical line that forms the arithmetic average of all these peaks and valleys. Then the surface roughness is specified as the difference between this and the average center line between the peaks and valleys. Typical values for as-fired, 96%, unglazed alumina surfaces are 20 to 30 microinches. For glazed alumina and sapphire the values are around 1 microinch.

Surface lapping and polishing can be used to improve the surface finish specifications. All that this does is to grind off the peaks that stick up with a diamond wheel, leaving a series of flat mesas with valleys between them. This procedure is both expensive and unnecessary for normal thick film use—and the end results may be less desirable than the as-fired surface. The grinding can actually worsen the surface finish. It can pull out whole crystals and grains from the surface, creating large voids.

Glaze. Manufacturers make available a choice of two surface finishes—glazed and unglazed. The glazing technique is simply the addition of a sprayed, dipped, or screened-on coat of low-temperature glass that is fired onto the ceramic surface to fill in the voids and smooth out the surface. "Unglazed" refers to the natural or "as-fired" surface.

Glazing does produce a smoother surface, and for this reason it is generally used for thin film work, where metal and chemical film deposition is involved. For good deposition the surface must be as smooth as possible. However, for thick film work, the unglazed surface is adequate, and as mentioned above, the extra roughness does actually help the fired film adhesion. Due to the poor thermal conductivity of the coating of glass, glazed surfaces are not able to dissipate as much heat as unglazed ones. Glazed substrates are more expensive, a possible disadvantage when large quantities are involved.

Glazing seals the surface of the substrate, stopping up all the pores. This could be used to advantage if the circuit is to encounter very humid conditions. Water absorbed in the ceramic pores could, in some cases, lower the surface resistance of the substrate and interfere with circuit operation. Glazing would seal this surface and eliminate the water absorption.

Camber. One important surface condition is referred to as warp or camber. As will be seen later on, this can have quite serious effects on flip-chip and beam-lead bonding. This can also create uneven paste deposition problems in the printing step and, if vacuum platforms are in use, causes the vacuum to leak around the edges. When a heated stage is being used, such as for die bonding or for thermal compression wire bonding, a warped substrate will not be pulled down into intimate contact with the heated surface at all points, leading to a substrate temperature differential that will provide inconsistent bonding results.

Camber is specified in mils per inch. Typical values are less than 4 mils per inch, a deceptive figure since it usually refers to the overall camber on a substrate. In other words, the measuring points are one inch apart. Between these measuring points, some large variations can occur. If the distance is measured only over the area where the variation occurs, and this figure multiplied by a factor to convert it to a one-inch-apart measuring method and expressed as mils per inch, the camber figure can be quite amazing. This "local" camber can run up to 10 or 15 mils per inch and if a flip-chip device were bonded in this area, the bonds would not be equally bonded. This can seriously degrade bond strength and reliability. Again, lapping or grinding can be used to improve the camber by removing the high areas, but are expensive procedures and may produce undesired effects on surface roughness.

To sum up, 96% unglazed alumina substrates are the workhorse of the thick film industry. Different materials may be used for different requirements, such as beryllia for high-power dissipation. Glazing will provide a smoother finish, but usually this is not desirable. Wherever possible, holes should be designed out and standard sizes used as much as possible to keep the costs down on the raw substrates.

5 — Screen Printing

Thick film hybrid circuitry starts with metallization patterns screen-printed onto a ceramic substrate to form conductors and specially formulated resistive pastes screened on to form resistors.

Silk screening, or screen-process printing as it is more properly called these days, is a technique that has been known for hundreds of years. It is only relatively recently that it has been applied to the field of electronics, although it has been used for some years in the manufacture of printed circuit boards.

As the name implies, a silk screen is a mesh of finely woven silk or synthetic cloth, used as the support for a printing pattern that blocks out certain areas of the mesh. Paste is forced through the open areas of the screen onto the desired surface by passing an ink-loaded rubber blade, called the squeegee, across the screen. The screen is slightly above the surface to be printed, rather than actually being in contact with it. When the squeegee is passed across the screen, it deflects the screen downwards to contact the printing surface and leave a deposit of paste. As the squeegee passes across the screen, the screen pattern comes back up to its original position after that particular section has done its printing job.

The paste used to deposit the patterns has a high viscosity. It will not flow freely through the tiny mesh openings in the screen, but has to be forced through by the action of the squeegee moving across the screen. Paste screening for thick film circuits has become somewhat more sophisticated than in the old days of a natural silk screen, a wooden frame, a knife-cut stencil, and a hand squeegee. For thick film hybrid work, the silk has almost exclusively been replaced by a woven wire mesh made of stainless steel. Stainless steel has several advantages. It is more evenly woven and stronger than synthetic materials. It does not sag and stretch like the synthetics. It will not unravel tiny threads that start to blur the printed pattern after use, as silk will. The paste slips through the metal mesh more easily than it will through an organic fiber. The wooden frame used to support the printing mesh has been replaced by cast and machined frames, made of cast aluminum or rigid epoxy and constructed to precise tolerances. Wooden frames are occasionally still used; however they tend to warp and bend quite easily, and they absorb some of the paste and solvents used in printing. For any kind of precise work, machined frames with stainless steel mesh are essential.

The actual process of screening the resistors and conductors has been reduced to a semiautomatic or automatic process to try and eliminate as many of the printing variables as possible. A thick film screen printer is a precision instrument, designed to eliminate the variables of hand screening. Screen-printed resistors have to be precisely accurate in film thickness and size for repeatable results. Forgetting variations in paste and substrates for a moment, conductor and resistor dimensions can be altered by:

1. The rate of moving the squeegee across the screen.
2. The degree of parallelism between the screen and the substrate.
3. The height of the screen above the substrate or "breakaway."
4. The pressure of the squeegee on the screen.
5. The rate of ink feed.
6. The mesh count of the screen.
7. The emulsion thickness of the screen stencil.
8. The hardness of the squeegee material.
9. The degree of squeegee wear.
10. The angle at which the squeegee is used.
11. The tension of the screen mesh.
12. The direction of squeegee travel.

Moving the squeegee by hand across the screen is not precise enough to exactly duplicate all these parameters at every pass across the screen. It is essential to use a precise printing machine that can control as many of these variables as possible. Further variables in resistor-making creep in due to paste viscosity, paste batch, firing and drying processes, and other subtle degrading factors.

Mechanical screen printers are designed to bring the squeegee at a fixed, constant rate across the screen at a fixed angle with a fixed pressure, feeding the paste at a constant rate to the squeegee. The screen and substrate are kept in fixed relative positions during printing. The technique of producing repeatable resistors will depend, in part, on the ability of the machine to duplicate these conditions for substrate after substrate.

The Screen

The screen itself consists of a woven wire mesh of stainless steel stretched on a precise frame. A close-up of screen mesh is shown in Fig. 5-1.

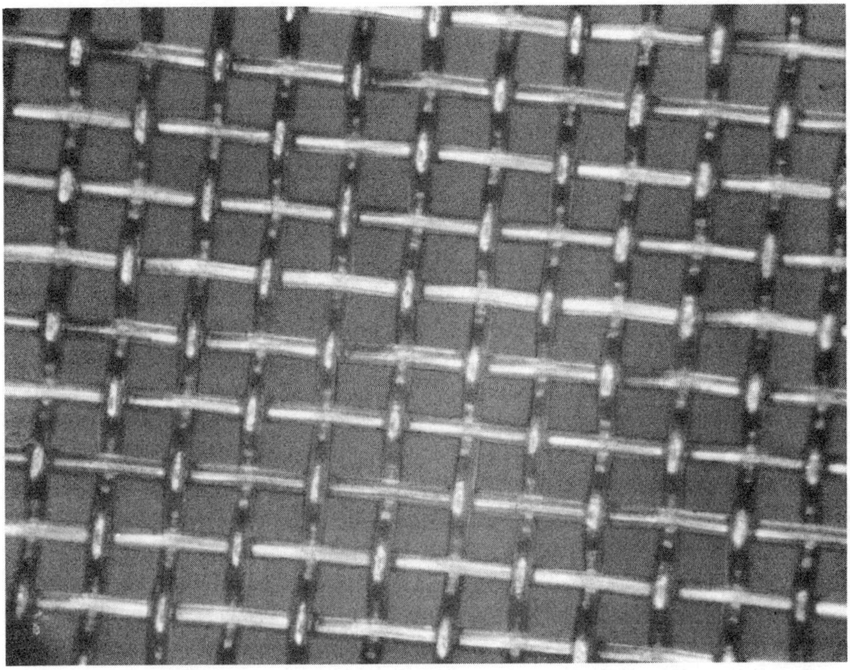

Fig. 5-1. Close-up of a stainless steel screen mesh before the stencil material is applied.

Although it is perfectly possible to stretch and attach the mesh to the frames in-plant, it is not really worth the effort. Although it is done every day by people in the screen-printing trades, dealing with microcircuit screens is a very different proposition from screening posters or textiles. It is very difficult to get even tension in the wire cloth used for printing thick film patterns and place it in the correct alignment on the frame. Current methods use a pneumatic or hydraulic machine to stretch the wire cloth from each side simultaneously. Then the frame is brought up and adhered with heat, epoxy bonding, mechanical devices, or a combination of all of these. Screens that are properly tensioned and ready for the emulsion can be bought for under \$10. This is a small price to pay for the work that goes into them.

Screen cloth comes in various mesh counts. This refers to the number of wires per inch of the mesh. Common mesh counts for thick film work are: 105, 150, 165, 180, 200, 230, 250, 270, 325, and 400. The mesh count of the screen to be used will depend on the pattern to be screened. If a small mesh (high mesh count) is used, a finer pattern with better detail can be resolved than if a coarse mesh (low mesh count) is used. The stencil or pattern made on the high-mesh-count screens has more wires to adhere to and produce the fine detail. But because there are more wires per inch, the open area, or area between the wires, is smaller. Less paste is forced through onto the substrate, resulting in a thinner deposited film. Since the thickness of resistor films is an important determining factor in the value of resistors, very fine mesh screens are generally not used for resistor printing. When printing conductors, where thickness is not as critical and fine detail can make the difference between being able to make the circuit or not, fine mesh is often used. Many people use 250 or 325 mesh for their conductors. For resistors, a coarser mesh should be used to ensure that an adequate film thickness is deposited; 165- and 200-mesh screens are common for resistor screening.

The Stencil

Before using the screen, a pattern has to be placed on the mesh to leave open the areas where paste should be printed through and to block out the areas where it should not. This pattern is called the screen stencil. There are two major classes of stencil: the emulsions and the metal masks. The classes break down as follows: direct emulsion, indirect emulsion, direct-indirect emulsion, direct metal mask, and indirect metal mask. Table 5-1 shows a comparison of the five classes.

Emulsion Stencils. The *direct emulsion* is probably the most common in use at this time. The direct emulsion consists of a liquid form of polyvinyl alcohol or polyvinyl acetate (available from graphic arts and screen-printing suppliers under the generic heading of "direct emulsion") that is made light-sensitive by adding a specific amount of sensitizer. Sensitizers used are

Table 5-1. Comparison of the Various Screen Stencil Types.

Stencil	Life Expectancy	Cost	Ease of Preparation	Turnaround
Direct emulsion	Good	Inexpensive	Simple	Very fast
Direct-indirect emulsion	Reasonable	Inexpensive	Simple	Very fast
Indirect emulsion	Poor	Inexpensive	Simple	Very fast
Indirect metal mask	Excellent	High	Complex	Long
Direct metal mask	Excellent	Highest	Very complex	Longest

water-based solutions of potassium bichromate or dichromate, sodium dichromate, or ammonium dichromate. When handling the sensitizer in either the powder or liquid form, it is wise to wear rubber gloves and avoid breathing any of the dust when mixing the sensitizer solution. The dichromate families are strong irritants and can cause serious skin and lung irritations and ulcers that can take a long time to heal. This is not intended to be a somber note—it is just good practice to be careful when using chemicals in general.

The sensitizer and the emulsion are mixed thoroughly, according to the manufacturer's instructions, and the resulting mixed emulsion is poured onto a prepared screen. The emulsion is spread to the desired thickness with a hand squeegee or other flat rubber blade. The squeegee blade is run across the outside surface of the screen to spread the emulsion evenly and to scrape off the excess. Then the blade is run over the inside of the screen to do the same. This is repeated several times, until a thin, even coating of emulsion is spread across the screen wires, with all the excess emulsion removed.

There is quite a bit of technique involved in applying a direct emulsion to a screen. It is easy enough to apply, but it takes a lot of practice to be able to repeatably apply an even coat of the same thickness. The emulsion thickness is one of the controlling factors in printed resistor thickness, which in turn is a factor in the final resistance value. Depending on the manufacturer's paste calibration techniques, the desired fired film thickness of the resistors will be between 0.6 and 1.0 mil. Practice and technique are required to repeatably make screens to produce this result. A close-up of a screen mesh with emulsion applied is shown in Fig. 5-2.

The next step is to dry the emulsion on the screen. Warm air blowing lightly across the screen can be used to speed this up. A second or third coating of emulsion, with drying between each coat, can be applied if a thicker stencil is desired. Exposure of the screen stencil is accomplished by placing the photographic pattern in intimate contact with the screen in the desired location and orientation, holding the two this way in a vacuum frame or sponge-rubber-filled frame, and exposing them to the light.

Artwork is almost always laid out on perpendicular, straight lines, so it is very important that the mesh be parallel to the artwork, or a phenomenon called "sawtoothing" or "stairstepping" is produced. An example of this is shown in Fig. 5-3. This occurs when the pattern and the mesh are not exactly matched and the edge of the pattern does not follow one wire exactly, but crosses one or two wires. When the stencil is exposed and developed, the pattern will have an unsupported side where it crosses the open areas between the meshes. With a direct emulsion this may or may not stay there when the emulsion is washed out after exposure. With an indirect emulsion, the edge will be there to begin with, but after use the unsupported edge will break down back to the nearest mesh. In either case, the unsupported edge will break down back to the nearest mesh eventually and what is left is a series of

Fig. 5-2. Close-up of a stainless steel mesh after direct emulsion is applied and dried. Stencil exposure has not yet taken place.

steps like the edge of a saw blade. Hence the names sawtoothing and stairstepping. The line that is printed by this screen will have a ragged edge and dimensional inaccuracy. The two points to keep in mind are that resistors should not be laid out at an angle to the mesh, but must be parallel to it, and that the artwork pattern must be aligned with the mesh before the screen is exposed.

The pattern used is a photographic positive. The dark areas on the photograph correspond to the areas that will be printed by the screen. For example, when screening resistors, the desired pattern will be black on the photograph and the open areas will be clear.

When exposed to light, as shown in Fig. 5-4, the sensitized and dried emulsion polymerizes to form long, hard molecular plastic chains. The areas that are not exposed, or are protected by the opaque photographic areas, will not polymerize. These unexposed areas can be washed out with a gentle spray of warm water, to form the open areas of the stencil. The exposed and polymerized areas will not be affected. This produces open areas of mesh where the paste is to be screened through, surrounded by areas blocked out by the emulsion. The plastic stencil is resistant to the pastes and solvents used in printing.

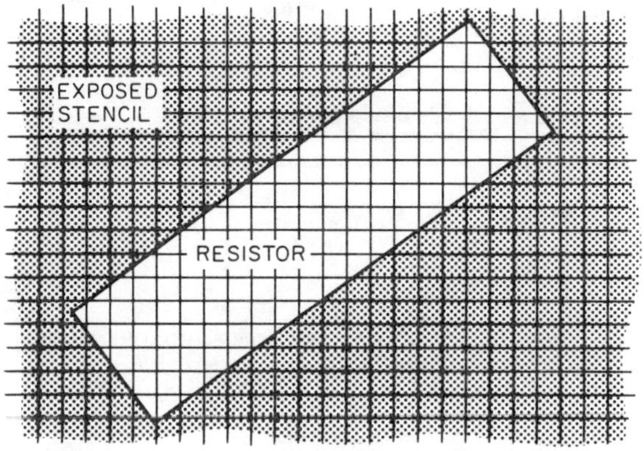

(A)

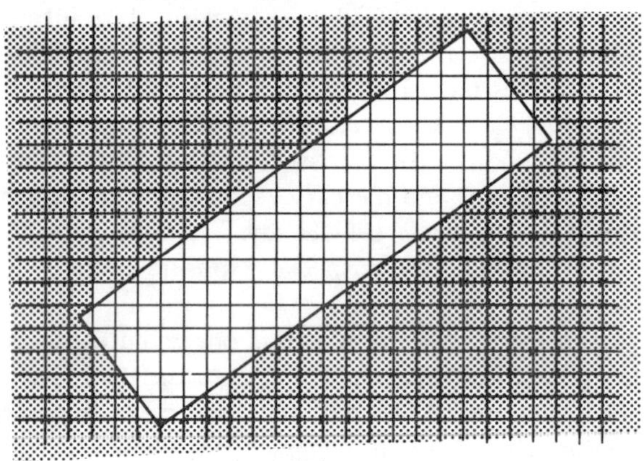

(B)

Fig. 5-3. Sawtoothing or stairstepping on screen stencils. (A) Exposed pattern on screen. Note that the stencil crosses between meshes and has unsupported edges. (B) Eventual breakdown of emulsion leads to pattern degradation. The pattern breaks down to the nearest mesh, causing distortion and preventing the printing of a true reproduction.

The term "light" was used several times in the foregoing description for screen exposure. This must be defined more clearly. Sensitized direct emulsion is exposed by ultraviolet light. The screen-making room, therefore, must be equipped with tungsten (ordinary incandescent) lamps. Yellow lights, or "bug" lights, are better, but with either white or yellow light the intensity should be kept fairly low when coating and drying screens, since very bright lights can start the polymerization reaction. Fluorescent lights will poly-

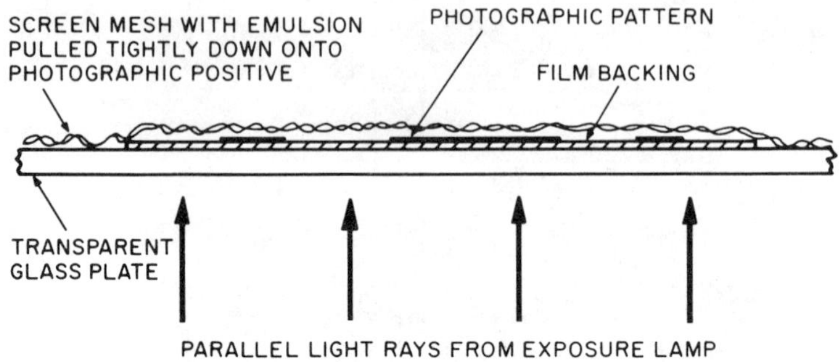

Fig. 5-4. Screen stencil exposure.

merize the emulsion, although yellow fluorescents can be used if necessary. Sunlight, of course, will expose the emulsion.

To expose the emulsion, any light source with a high ultraviolet content can be used. Arc lamps are the best, but other sources include quartz-iodine lamps, ultraviolet fluorescent lamps, white fluorescent lamps, sunlamps, and mercury vapor lamps. Careful precautions to shield the operator should be taken when using ultraviolet light sources. Shortwave ultraviolet light, such as that produced by arc lamps and mercury vapor lamps, causes cumulative and irreversible damage to the optic cells of the eyes in a very short time. It is wise to have safety interlocks and special exposure frames when using this type of equipment.

Exposure time will vary from about one minute to five minutes or longer, depending on the light source used and the desired emulsion thickness. The thicker the stencil desired, the longer the exposure time. For resolution of fine detail, a fairly thin stencil is desirable; an exposure time of two or three minutes may be adequate. For coarser work, a thicker stencil can be used, bearing in mind that this will result in a thicker deposited film. The only way to determine the optimum exposure time is by a series of trial-and-error exposures. Light sources and exposure fixtures vary too much in intensity and spectral density to make any definite rules.

After the exposure, the screen stencil is "developed" by washing it with a gentle spray of cold or warm water. If the wash-out water is too hot, it will soften the stencil and may distort the pattern. The spray should be directed to both sides of the screen and the pattern washed until it is clear. The unexposed emulsion must be washed completely off the screen, or it may run down into the pattern and block some of the open areas. The final step is to dry the screen, and it is ready for printing. A typical conductor pattern on stainless steel mesh is shown in Fig. 5-5.

The *indirect emulsion,* sometimes called transfer emulsion or transfer film, is similar to the direct emulsion, except that the sensitized emulsion,

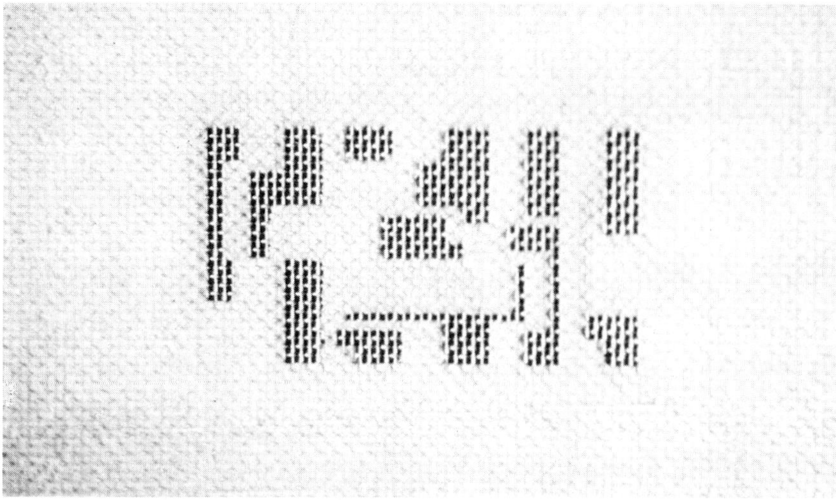

Fig. 5-5. Exposed and washed screen stencil on screen ready for printing. View is from substrate side of screen.

usually gelatin based, is already spread in a thin even film onto a backing sheet of paper or plastic. The emulsion is exposed and developed while it is still attached to the backing, before it is adhered to the screen. After the pattern has been washed out and while the stencil is still wet, it is smoothed onto the screen with firm, but gentle, pressure. After the stencil is dry, the backing is peeled away to leave the finished screen. Figure 5-6 compares the processing steps for direct and indirect emulsions.

The indirect method has two advantages over the direct emulsion. First, the emulsion comes already coated on a backing material and when it is transferred to the screen it will have an even thickness all over the screen. This thickness is repeatable from screen to screen whenever a new stencil is made, something which requires quite a bit of expertise to do with a direct emulsion. This apparent advantage could also be a disadvantage if emulsion thickness is being used to change the sheet resistance of a paste by altering the printed film thickness.

Another advantage of the indirect method is that the pattern is not as dependent on the mesh as a direct emulsion stencil. Direct emulsion will only go from mesh to mesh and will produce sawtoothing if the edge of the pattern does not coincide with the mesh. The indirect emulsion is already washed out before adhesion to the screen, and will go between meshes more readily. This feature is useful for resolving fine detail, making the indirect emulsion superior to the direct for fine printing.

These two advantages are overshadowed by a much greater disadvan-

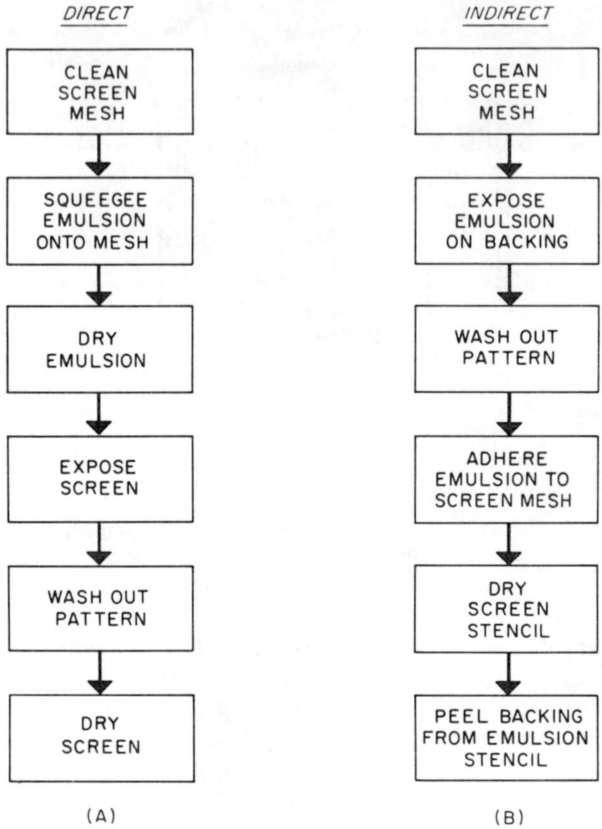

Fig. 5-6. Screen-stencil-making process steps. (A) Direct emulsion. (B) Indirect emulsion.

tage. The life of the indirect stencil is much shorter than a direct emulsion stencil. The adhesion to the mesh is much poorer for indirect emulsion and, with the additional stresses created when the emulsion covers only part of a mesh opening, the indirect stencil cracks, flakes, and generally breaks down much faster than a direct emulsion.

The characteristics of the indirect emulsion make it superior to direct emulsion for fine-line, short-run printing cycles, but for overall usage and long life, the direct emulsion is superior.

The *direct-indirect emulsion* technique, which is not generally used at the present time, borrows from the techniques of both direct and indirect stencil methods. This technique uses a plastic backing that is precoated with emulsion, to provide emulsion thickness control and repeatability. The emulsion is adhered and dried on the screen and the plastic backing removed before exposure and development takes place. This system is a little more complicated to use. As expected, it gives results between the direct and indirect characteristics.

Actual screen stencil life is very difficult to predict. It depends on the type of use the screen is given and no matter how carefully the screen is used, eventually there will be physical deterioration and breakdown of the screen emulsion. One of the major causes of screen deterioration is excessive squeegee pressure during the actual printing process, which occurs when the distance between the substrate and the screen is large or when the printing machine is misadjusted. The squeegee pressure has to be high to force the screen down into contact with the substrate, leading to excessive deformation of the screen. This causes the screen itself to change its mesh tension rapidly, resulting in stresses and cracking in the emulsion. The screen will also be shortlived if the squeegee pressure is too high, even if the screen-substrate distance is small. In this case, the screen will be forced below the substrate surface outside the pattern area, come up to the substrate surface over the pattern, and then drop back below the surface level at the other side of the pattern. Excessive wear and kinking of the screen will result at the corners of the substrate. This is called "coining." The sequence in coining a screen is shown in Fig. 5-7. Coining breaks down the screen material, as well as the emulsion. This problem can be reduced by having a recessed area, the size of the substrate, in the printing platform. If the surface of the substrate is only 4 or 5 mils above the printing platform, the screen will be supported by the printing platform and the coining will be reduced. A photograph of a badly coined screen is shown in Fig. 5-8.

The other major cause of short screen life is short printing runs. After each short run, the screen will be taken off and cleaned. This cleaning will very quickly result in the deterioration of the screen pattern.

With reasonable care, the direct emulsion can be expected to last from 10,000 to 20,000 passes in a continuous run. For shorter runs with cleaning after each, the life will be shorter. The indirect emulsion will last from about 1000 to 2000 passes.

Before an emulsion stencil can be put on a screen, the screen mesh itself must be properly prepared. New screens, straight from the manufacturer, cannot be used. They usually have a film of oil and dirt left over from the manufacturing process and are often covered with a film of light oil for protection of the mesh. If this film is not removed, the emulsion will not adhere to the mesh. There are several screen-cleaning preparations available from screen equipment suppliers. The individual cleaning processes vary from one plant to another. Processors use various mesh-cleaning methods, from using weak acid solutions to dissolve the contamination, to burning off the oil and dirt with a bunsen burner. Needless to say, some of these methods are quite hard on the screen and are not really recommended.

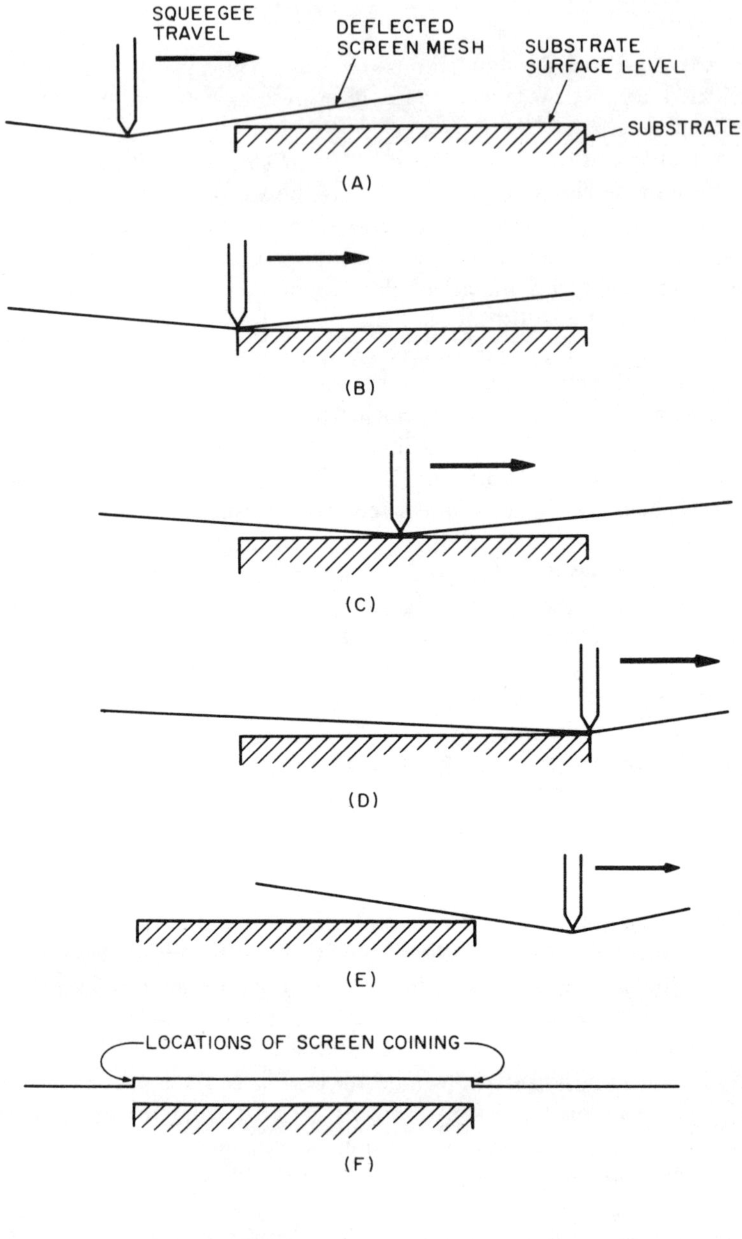

Fig. 5-7. Sequence of screen coining. (A) Squeegee approaches substrate and deflects screen below substrate surface level. (B) Squeegee comes to edge of substrate and has to rise up over the edge. Stress occurs at edge of substrate where contact is made. (C) Squeegee prints on substrate. (D) At far end of substrate, squeegee drops back below substrate surface level. Stresses again occur at edge of substrate where screen contact occurs. (E) Squeegee continues to move below substrate surface level until the end of squeegee stroke. (F) Eventually the screen will become coined at places where the edges of the substrate dig into the screen mesh from the motion of the squeegee.

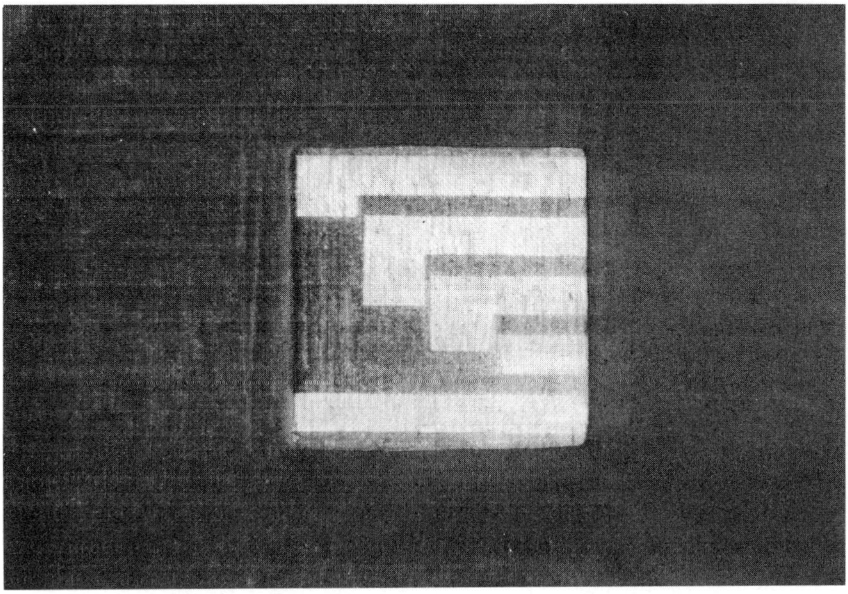

Fig. 5-8. A badly coined screen. Note indentation marks of substrate edges around screen pattern.

One of the best and simplest methods of screen cleaning is to use an abrasive household scouring powder, such as Ajax* or Comet.** These powders have the advantage of cleaning off the dirt and, at the same time, the abrasive action will roughen the screen wires slightly to provide a good surface for strong emulsion adhesion. The abrasive powder is sprinkled on the screen with a little water, then spread around with a sponge or paper towel, in a circular rubbing motion. Rubbing too hard will either stretch or kink the screen wires. A metal mesh screen that has a kink in the wire cannot be repaired and must be discarded. After scrubbing is complete, the screen is well rinsed in hot water with a strong spray nozzle. All the scouring powder and particles must be removed or the emulsion will not adhere.

Stainless steel screens may be re-used by cleaning off the old stencil. With careful treatment, the life of a stainless steel mesh can be quite long. Indirect emulsions should be cleaned off according to the particular manufacturer's recommendations. Hot water is often used to dissolve the emulsion.

Direct emulsions are washed off in a strong oxidizing solution such as hydrogen peroxide or sodium hypochlorite (used in household laundry bleaches, such as Clorox***). The solution is poured on both sides of the

* Ajax is a trademark of the Colgate-Palmolive Company.
** Comet is a trademark of Proctor & Gamble.
*** Clorox is a trademark of the Clorox Company.

stencil and allowed to soak for a few minutes. Then the screen is washed out with a strong spray of hot water. If aluminum screen frames are being used, care must be taken to try and keep the bleach off the frame, as it will corrode the aluminum.

Washing out a new stencil that did not turn out right is very easy, but washing out a stencil that has been used for some time can be quite difficult. Generally, the older the stencil, the harder it is to remove it. The pastes and solvents used for screening act as a barrier to prevent the stencil solvent (the bleach) from dissolving the stencil material. It is essential to clean the paste residues off the screen completely, before trying to clean off the stencil. Alternate washings of paste solvent and stencil solvent may be necessary in tough cases. The most common reason for having difficulty in washing out an old stencil is that solvents or pastes have not been cleaned off the screen well enough and block out the bleach.

Screen tension and deflection are important factors in the repeatability of a printed pattern. The tension of the stainless steel screen mesh is not a constant, but will decrease throughout the life of the screen. It is measured by pressing on the middle of a screen with a rounded object, and measuring the amount of force needed to deflect the mesh a certain distance, as is shown in Fig. 5-9. Special fixtures and gauges are available for measuring screen tension. For best results, screen deflection should stay within the limits determined by the manufacturer. Screen tension is an important factor in the screen-printing process. If too much force is required to deflect the screen, the mesh will stretch during printing and cause the stencil itself to distort. If the screen does not require much force for deflection, the mesh is too loose and will either slide slightly on the substrate during printing or will not snap up away from the substrate after the squeegee pass, both of which will cause the printed pattern to smear and lose definition.

The screen tension will change as the screen is used. A coarse mesh screen (low mesh count) will last longer than a fine mesh screen. A higher squeegee pressure has to be exerted on the fine mesh screen to force the paste through the smaller mesh openings. To screen resistors and conductors, it is advisable to use the coarsest mesh possible that is consistent with desired line definition and film thickness. The size of the screen frame also affects the mesh tension. An 8″ x 10″ screen will deflect more than a 5″ x 5″ or 4″ x 5″ with the same mesh count. Screen suppliers can supply information on recommended deflection and tension limits for the various meshes and screen sizes that they supply.

The size of the screen frame used depends primarily on the size of the artwork pattern to be used. The size of the screened pattern cannot be the same size as the screen. For good performance, a generous border should be left between the pattern and the sides of the frame. A good rule of thumb is to use a size screen that leaves at least a 1-inch border between the pattern

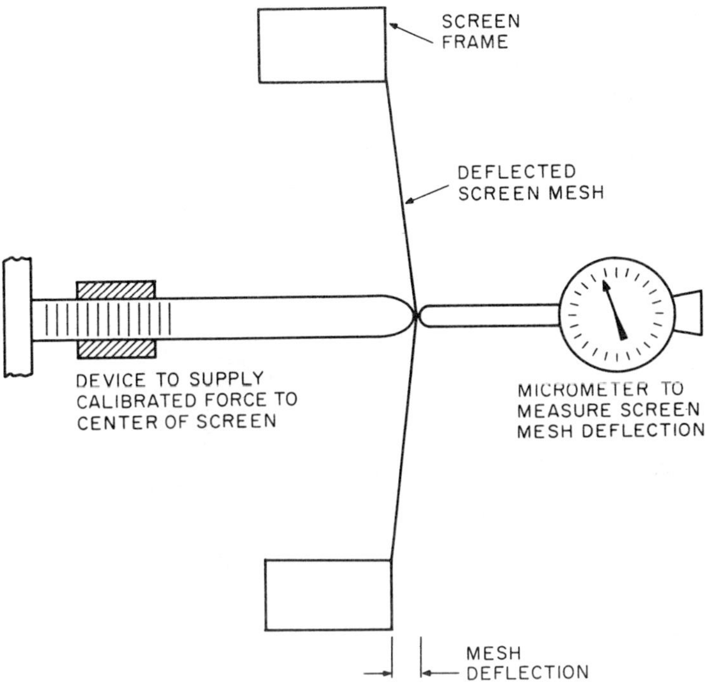

Fig. 5-9. Measuring screen mesh tension.

and the frame. For example, a 4″ x 5″ screen would be good for printing patterns up to about 2″ x 3″. This border is to allow the screen to deflect evenly down to the substrate as the squeegee moves across the pattern. If the pattern is the same size as the screen, uneven stresses and tensions will develop around the edges of the pattern. This will accelerate screen and stencil deterioration and cause poor printing definition at the edges of the pattern.

Metal Mask Stencils. The second major class of screen stencils is the metal masks, which also come in the two subcategories of direct and indirect. The *indirect metal mask* is one step above the direct emulsion method. However, it is also similar to a transfer stencil. A cross section of an indirect metal mask is shown in Fig. 5-10(A). The pattern is formed by etching the open areas of the pattern into a precisely formed metal foil, usually 1 or 2 mils thick. Stainless steel, molybdenum, and Kovar* (iron, cobalt, nickel alloy) are commonly used foil materials. The etched foil is bonded to the bottom of a stainless steel mesh screen. This type of stencil has the fine detail

* Kovar is a trademark of Westinghouse Corporation.

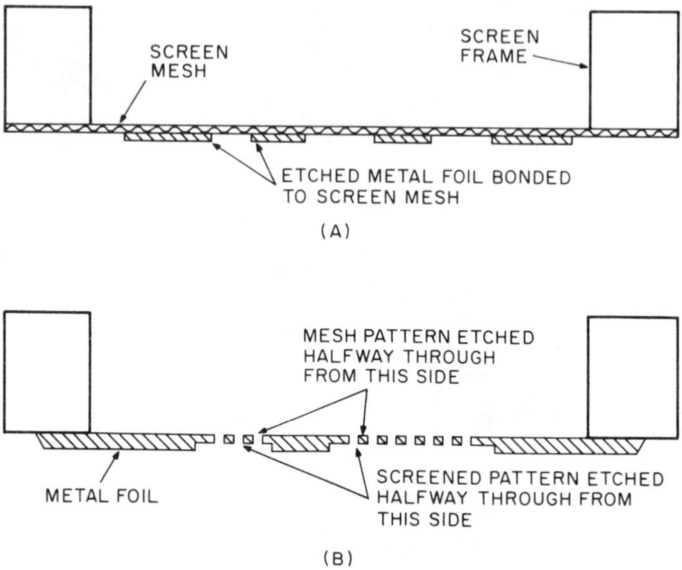

Fig. 5-10. Cross section of (A) an indirect metal mask screen and (B) a direct metal mask screen.

capabilities and thickness control of the indirect emulsion, and being a metal foil, it has a very long life. By the same token, it can have a very short life if mistreated. Anyone who has tried to straighten out the creases in a sheet of aluminum foil will realize that any dents or kinks will ruin the screen. The etching and bonding processes of metal masks are quite complex, and are better left to out-of-plant vendors who specialize in such items.

The *direct metal mask* is a further refinement on the indirect metal mask. For greater precision and accuracy, direct metal masks do away with the stainless steel mesh and use only an etched foil. A cross section of a direct metal mask is shown in Fig. 5-10(B). The foil used is 3 or 4 mils or even more, depending on the application. The printing pattern is etched into the foil on one side about halfway through. A series of holes, to take the place of the mesh, is etched through from the other side to provide a passage for the paste. This mesh pattern is individually generated for the particular artwork layout used and may vary from one part of the layout to another, depending on what type of holes best fit the pattern. This eliminates the effects of stairstepping and sawtoothing and other partially covered mesh effects by eliminating the mesh dependance. Direct metal mask screens are capable of extremely fine line printing.

The use of metal mask screens brings up the subject of contact and off-contact printing. Most thick film printing is done off-contact, where a gap is left between the screen and the substrate. The squeegee deflects the screen down to the substrate as it passes over the pattern. Off-contact printing is done with a space of about 10 to 60 mils between the screen and substrate, depending on other printing parameters. This is called the snapback or breakaway distance. To improve the quality of the print, and for fine detail work, printing is also done with the screen coming down right next to the substrate, before the squeegee pass is made. This is called contact or on-contact printing. In practice, actual contact between the substrate and screen is not made, but a slight gap of 1 or 2 mils is used to account for variations in substrate camber and thickness. A metal mask will deflect very little, so it is usually used in the contact mode.

Considering the printing advantages offered by direct metal masks, it would appear that this is the type to use—but not necessarily. The main deterrent is price. The metal masks are expensive to make, are done out-of-plant by experts with sophisticated equipment, and involve delays in turn-around. Once a metal mask is made up, it cannot be changed if the circuit requires modifications—a new mask has to be made. On the other hand, the direct emulsion method is very cheap, and is quick and easy to do in-plant. Storage of the unsensitized emulsion does not present any particular problem, though it should not be subjected to temperature extremes, especially freezing. One trade-off to be considered might be direct emulsion for proto-typing, quick turnaround, and short production runs—then a metal mask for long runs, or fixed pattern production runs.

Squeegee Material

The selection of a squeegee material is an important consideration and a subjective one. Some processors find that one type works very well for them; others find that they can only get good results out of another type. The basic considerations for squeegee selection are that the material is chemically resistant to the many solvents, pastes, and other chemicals that it will come into contact with, and also that it be mechanically resistant to the constant abrasion it will receive during the printing cycle.

There are three materials in general use: neoprene, polyurethane, and fluorocarbons or fluoroelastomers.

Neoprene is the best known of the squeegee materials, but it is being replaced in many plants by polyurethane. Neoprene wears quite badly, flaking pieces of rubber off into the paste and onto the screen. Neoprene also does not stand up to the solvents used in printing, and will start to swell and deform quite rapidly. The fluorocarbons are very similar to polyurethane, except the abrasion resistance is not quite as good and the cost is higher.

Squeegees are available in several different durometers or hardnesses. Durometers from about 30 to 90 are common. The softer materials (low durometer) do the best job of depositing an even film of paste, and they can tolerate relatively large variations in printing surfaces. However, they tend to wear very rapidly, thus altering their printing characteristics. The harder materials wear very well, but do not conform to substrate variations or print as efficiently as the soft materials. Usually a compromise is made. Most people use a medium durometer (around 60) for reasonable wear as well as for getting good printing characteristics.

Squeegees must be kept sharp for printing good detail. The sharper they are, the smaller is the portion of the screen that is actually in contact with the substrate at any one time during printing, thus increasing fine printing capabilities and reducing the possibility of smearing. As the squeegee wears round, the fine detail becomes poorer and the deposited film thickness becomes greater. Depending on the type of work being done, it may be necessary to break in the squeegee by printing several hundred substrates before going to actual production. This will eliminate the variations that occur during early wear and seating of the squeegee. Squeegees are cheap and can be discarded when they lose their sharpness. Equipment for resharpening some types of squeegees is also available.

Printing

Thick film screen printing machines are precise and complex pieces of equipment. Each brand of machine is different, so only a few generalities can be observed:

1. The substrate should be held flat and securely on the printing platform.
2. The substrate surface must be held parallel to the screen.
3. Screen-to-substrate spacing should be adjustable and repeatable.
4. Squeegee speed should be adjustable and constant at each adjustment.
5. Squeegee pressure should be adjustable and precisely controlled.
6. Substrate positioning should be adjustable in the x, y, and circular directions.
7. Substrate edges should be held parallel to the pattern and have a precise locating method for repeatability.

To start the printing cycle, a substrate is placed on the printing platform. The substrate is positioned either by locating pins or by placing it in a recessed area in the printing platform. Recessing the substrate into the printing plate will help prevent coining the screen, as mentioned previously. The top of the substrate should be about 5 mils above the platform to allow for substrate variations. It is held down on the platform by vacuum holes under the plate. Next, the screen is brought to the desired printing distance

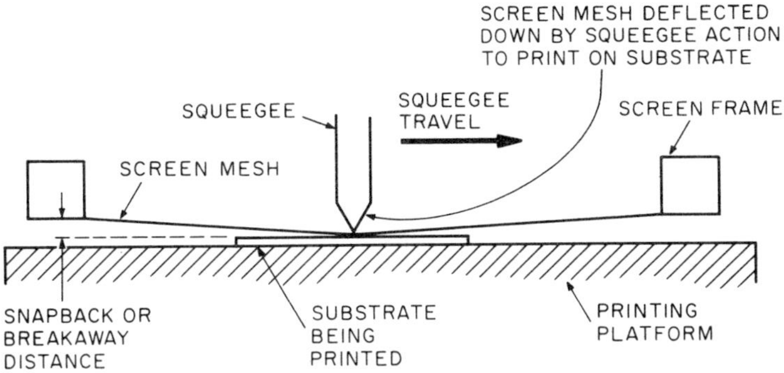

Fig. 5-11. The screen printing cycle.

from the substrate. Either the screen is fastened in a printing head that lowers or moves sideways over the stationary substrate, or the printing platform rises or moves sideways under the stationary printing head. In either case, the effect is the same. A common snapback spacing is 15 to 30 mils. The paste-loaded squeegee is brought across the screen, deflecting it and depositing paste through the pattern onto the substrate. This is shown in Fig. 5-11. The speed at which the squeegee moves across the screen is important, but not critical. If it moves too fast, air is trapped in the paste as it is forced through the mesh; if it moves too slowly, the screen stays in contact with the substrate too long, leading to smearing and loss of fine detail. The speed should be kept between about 4 inches/second and 8 inches/second. When the squeegee has completed its cycle, either the printing head or the printing platform moves in such a way that the printed substrate can be unloaded and a fresh one loaded. Automatic machines with shuttle feeds and unloading are available for high-speed printing.

Once the printing run has been started, it should not be interrupted. If the screen is left with paste on it for more than a few minutes, the solvents will evaporate and start to clog the screen, necessitating that printing be stopped while the screen is cleaned. If the run has to be stopped, it is best to clean the machine and screen and start all over again.

Squeegee pressure is directly related to screen life, as discussed earlier. Squeegee pressure should be kept as low as possible while still printing a good image. Typically, the squeegee is adjusted during machine setup to just touch the top of the substrate, then adjusted to come a certain distance further on down below the surface of the substrate during actual printing. Five to 25 mils overpressure or distance below the substrate contact point are common. Other methods use weights and mechanical methods to produce a pressure calibrated in pounds. Different machines are set up in different ways—follow the manufacturer's instructions for the machine in use.

Fig. 5-12. Gold conductor pattern.

After printing the pattern, the substrates should be allowed to sit in a dust-free environment at room temperature to allow the paste to level. This can take from a few seconds to 15 minutes, depending on the paste viscosity and pattern geometry. This step allows the paste to level and even out the impressions of the screen mesh.

Following leveling, the substrates go through a drying cycle. This drives the volatile solvents out of the paste and prepares the printed pattern for the firing cycle. Drying the substrates like this hardens the printed film right away and prevents the paste impressions from settling further and flowing on the substrate. The paste solvents are removed before the substrates go into the firing furnace, eliminating a source of atmospheric contamination in the furnace. Drying can be done on a hot plate, in a batch oven, on a conveyor dryer, and with infrared dryers. Drying temperatures are usually about 80° - 125°C and the substrates are held at that temperature for 10 to 20 minutes. If the temperature is much higher than 125°C, the binders in the paste may start to soften and plastic flow take place. Manufacturers' recommendations will vary, depending on the pastes used. Infrared lamps for drying have the advantage of drying the printed film all the way through. Other methods of drying can cause a dry crust to form on the top of the paste first, making it difficult for the solvents still trapped underneath to dry out completely. For fast processing, the dryer may be incorporated onto the inlet end of the furnace. The wet, just-printed substrates are placed right onto the furnace belt from the printer so that the patterns can dry and then go right on into the furnace for firing.

After drying, the printed film will be relatively hard and sturdy. The substrates can be stored in a dust-free container for firing later, or, if other patterns are to be printed on top, this can then be done. When the final printing and drying is done, the substrates are ready for firing. Figure 5-12 shows a printed conductor pattern.

When printing is finished, the screen and the squeegee should be carefully cleaned along with any other parts of the machine that come into contact with the pastes. Solvents recommended by the paste manufacturer should be used. If none is recommended, there are various satisfactory solvents such as acetone, butyl cellosolve,* methyl ethyl ketone, and trichloroethylene. These solvents should all be used with adequate ventilation or in a fume hood, and care should be taken with sparks or open flames. Since skin contact with solvents should be avoided, a wise safeguard is to wear rubber gloves as well as safety glasses or some other form of eye protection.

* Cellosolve is a trademark of Union Carbide Corporation.

6 — Thick Film Pastes

Thick film conductors, resistors, and dielectrics are deposited on the substrate by screen process printing. All three come in the same form—a paste. These pastes have high viscosity and will not flow through the screen mesh by themselves, but have to be forced through the wires by the action of the squeegee. Paste viscosity is very important. If the viscosity is too low, the paste will run and smear the pattern; if the viscosity is too high, either the paste will not print through the screen, the printed pattern will be too thin, or the print marks from the screen will not level out after printing, leaving a series of ridges in the screened pattern.

The pastes in general use for thick film circuits are called cermet pastes, the name being a combination of the words ceramic and metal. The pastes consist of a blend of three basic components: (1) the base material for the desired characteristics, either conductive, resistive, or insulating; (2) an organic binder; and (3) a volatile solvent to control the viscosity.

When the pastes are received, they should be checked. This would include a viscosity check and a sample screening and firing test. Special viscometers (viscosometers) that use only a very small sample are available for paste viscosity measurements. The paste should be vigorously stirred or blended to make sure that all the solids are in suspension and evenly distributed in the binder. This can be accomplished by vigorous hand-stirring with a stainless steel spatula or by using a laboratory blender. All this may alter the viscosity of the paste quite drastically, so store the pastes for about 24 hours before performing the viscosity checks. Many pastes require storage on a jar roller mill. This is a series of slowly rotating roller storage racks that keep the paste jars constantly rolling slowly on their axis to prevent the solids from settling out again. Roller speeds of around 1 to 10 revolutions per hour are used. Many pastes are available that do not need roller storage, but only need hand stirring before use. Again, follow the manufacturer's recommendations.

When the viscosity is measured, it should be recorded and checked to see if it meets the desired specifications. If the viscosity is too low, it will increase as solvents evaporate to the air, when the top is off the jar. If the viscosity is too high (as it will eventually become when the solvents evaporate through exposure to the air during use), then solvents can be added to lower the viscosity. Use the solvent recommended by the manufacturer, if this information is available. If it is not, the following are suitable for most pastes: technical pine oil or butyl carbitol* acetate (BCA).

It must be kept in mind that the characteristics given in the following sections are general guidelines only. The final characteristics of a particular paste rest with the particular manufacturer. For example, a manufacturer may have different types of platinum-gold conductors, some of which are directly die bondable while others are not. One must follow the manufacturer's recommendations and specifications for the particular paste in use.

Conductors

In discrete component circuitry, the various electronic parts are connected by point-to-point wiring. Printed circuit boards use copper tracks to connect one component to another. Similarly, thick film circuits use interconnects formed by conductive pastes that are screen printed onto a ceramic substrate, then fired at high temperatures to form conductive paths. The conductive part of the pastes consists of the noble metals, gold, silver, platinum, and palladium, used singly and in combination. There are some other systems in use, such as copper-based pastes and screenable epoxies, but since they are not in general use for making thick film conductors, none of them are discussed here.

The combinations used for conductor pastes are silver (Ag), palladium-silver (Pd-Ag), gold (Au), palladium-gold (Pd-Au), and platinum-gold (Pt-Au). Some of their characteristics follow. The more important considerations are summed up in Table 6-1.

Silver. This was the conductor used when the first screen-printed hybrid circuits were being made. Today it is not generally used, except for some low-cost, high-volume applications. The main problem encountered with pure silver is the migration of silver ions. This only occurs in the presence of moisture. If the package is hermetically sealed, or if it does not meet with any moisture, migration will not occur and silver may be safely used. Silver has the highest conductivity of the conductor pastes and it is easily solderable. However, resolderability is poor and silver conductors are not directly die-bondable. A directly die-bondable conductor can have transistor dice directly bonded to it through the application of heat and a scrubbing action. See

* Carbitol is a trademark of Union Carbide Corporation.

Table 6-1. Conductor Properties.

	Silver	Palladium-Silver	Gold	Palladium-Gold	Platinum-Gold
Chemical symbol	Ag	Pd-Ag	Au	Pd-Au	Pt-Au
Sheet resistance values (ohms/square)	.001-.005	.02-.04	.003-.02	.02-.05	.015-.05
Solderability	Yes	Yes	No	Yes	Yes
Die bondability	With preform	With preform	Eutectic	Eutectic	Eutectic
Ultrasonic wire bonding	Yes	Yes	Yes	Yes	Yes
Thermal compression wire bonding	Yes	Yes	Yes	Yes	Yes
Relative cost	Least expensive	In-expensive	Expensive	Expensive	Most expensive

Chapter 10 for further details. A conductor that is not directly die-bondable uses more complex methods of die attach, such as conductive epoxies or the use of solder preforms. Speed and ease of die attachment are the considerations.

Lead-tin solders can be used on silver conductors, though a solder that contains 2% or 3% of silver will give better results. The silver in the solder will prevent leaching of the silver conductors. Solder obtains its good joint-making characteristics by dissolving some of the metals that it is joining together. This forms a very strong intermetallic bond, rather than just "glueing" the two metals together. Silver dissolves very easily in tin and, if care is not taken, the silver conductors will all dissolve off into the solder, and leave no pads to be soldered. The silver will continue to dissolve until the tin is saturated. The rate of solder dissolution can be slowed down by adding 2-3% silver to the solder, thus saturating the tin to begin with. Another way of controlling this leaching problem is to use a solder that has a high lead and a low tin content. This can be quite difficult to work with though, since the melting point is much higher than for ordinary 60-40 solder and some experience in high-temperature soldering will be necessary. Using a solder containing silver is a better method. Silver oxidizes quite rapidly upon exposure to the air, and if the circuit has been stored for some time before soldering, the conductors may have to be burnished or polished to remove the oxidation and provide a clean, solderable surface.

Palladium-Silver. This is the workhorse of the thick film industry. It was developed to try and overcome some of the shortcomings of the silver system, yet retain the low cost of silver. Palladium-silver is easily solderable, but is not directly die-bondable. Wire bonding is usually not a problem, but under certain environmental conditions, palladium-silver conductors can develop a thick oxide layer. This will cause difficulties with ultrasonic wire bonding, since the wire cannot scrub through this thick coat of palladium oxide. Oxidation is seen as a change in conductor color. Freshly fired palladium-silver is a bright silvery color, but as it oxidizes, it picks up a dark, dull film. Burnishing or wire brushing may be advisable before use, for circuits that have been stored for any length of time.

Direct eutectic bonding cannot be done with palladium-silver conductors. A solder preform of gold with 2-4% silicon can be used or the chips can be directly bonded to areas that have been presoldered. Generally lead-tin solder would not be used, because of a gold-leaching problem that occurs with gold-backed chips. This is the same as the silver-leaching problem mentioned above. Gold also dissolves very readily in tin. A low tin-content solder (95% lead, 5% tin) can be used if the cycle times are short to prevent leaching. Another way around the die bonding problem is to overscreen gold conductors just onto the areas where the chips will go. Then eutectic bonding can be used. This method has the disadvantage of requiring an additional screening step to deposit the gold.

Gold. A more expensive conductor material, gold is nevertheless very commonly used. Direct eutectic die attach is excellent and gold is readily bondable with both ultrasonic and thermal compression techniques. Because of leaching problems with the solder, gold is not generally considered solderable. As with silver, gold will readily dissolve off the substrate into the solder. Instead of lead-tin solders, gold-tin or gold-silicon solders can be used quite successfully.

Palladium-Gold. Formulated to make a more versatile conductor than the plain gold, most of the palladium-golds are solderable and demonstrate good thermal compression and ultrasonic bonding. Chips (either gold-backed or plain silicon) can be eutectically bonded to palladium-gold.

Platinum-Gold. This combination preceded the palladium-golds in an effort to improve the gold conductor. It has the same general characteristics as palladium-gold; the leach resistance to molten solder is superior to the palladium-golds, and the oxidation problems sometimes encountered with palladium-gold are far less likely with platinum-gold. Since platinum is more expensive than palladium, platinum-gold is the most expensive conductor material.

Most important in choosing a conductor is to choose one that is compatible with the resistor system used. The resistor paste manufacturer will usually recommend one or will give data on the type of terminations used for calibration of the paste. If the resistors and conductors are not metallurgically compatible, odd and unpredictable results will take place. Gold and silver migration and leaching problems have occurred, with ions moving out of the conductors and into the resistors and vice versa. This will create shifts in sheet resistance. It is generally advisable to stick with one manufacturer for both resistor and conductor pastes, as he will have tested the compatibility problems for his system. Yet, resistors from one company and conductors from another can be used quite successfully, and some processors feel that this is the way to get the best from both. Before doing this, it is very important to run careful tests for compatibility when deciding on a system. Following the resistor-conductor compatibility choice, other characteristics such as solderability and die bonding can be evaluated.

Figures 6-1 and 6-2 show a comparison between two resistor-conductor interfaces. Figure 6-1 shows a good resistor-conductor termination. The edges of the printed areas are even and well defined, and the transition is smooth. Figure 6-2 shows a resistor-conductor problem. Large bubbles have formed at the interface. The photograph also shows some overfiring of the resistors.

When choosing a conductor, there is one other effect that is sometimes overlooked. Conductor pastes have very low resistance, generally several hundredths of an ohm per square. Long paths of narrow conductors will add some resistance to the circuit. This can be a severe problem when dealing with low value resistors to begin with; any conductor resistance will affect the circuit. Severe misoperation of circuits has been traced to high resistance conductors. It is an easy thing to overlook, but an ohm or two in the right place may do odd things. The solution, of course, is to keep conductor paths as wide as possible, to create the least number of squares in the conductor length. A low ohms-per-square conductor is essential too.

Resistors

Screen-printed resistors are the trickiest part of manufacturing a thick film circuit. Anyone who has tried to make this type of resistor will say that whatever can go wrong, will go wrong. This is not meant to discourage beginners. It simply means that resistor processing has many, many variables that can creep into the system to alter the results, and that extremely close control of the process is essential.

To start at the beginning, resistor pastes are not easy to make. Remember that the paste manufacturers have their process problems too. It is extremely difficult to make one batch of resistive paste identical to another batch, due to variations in the particle size, solids content, and many other

Fig. 6-1. Close-up of resistor-conductor termination. Resistor is black area and conductor is gray. Note that the resistor overlaps the conductor, and that the conductor is wider than the resistor to ensure good contact between the two.

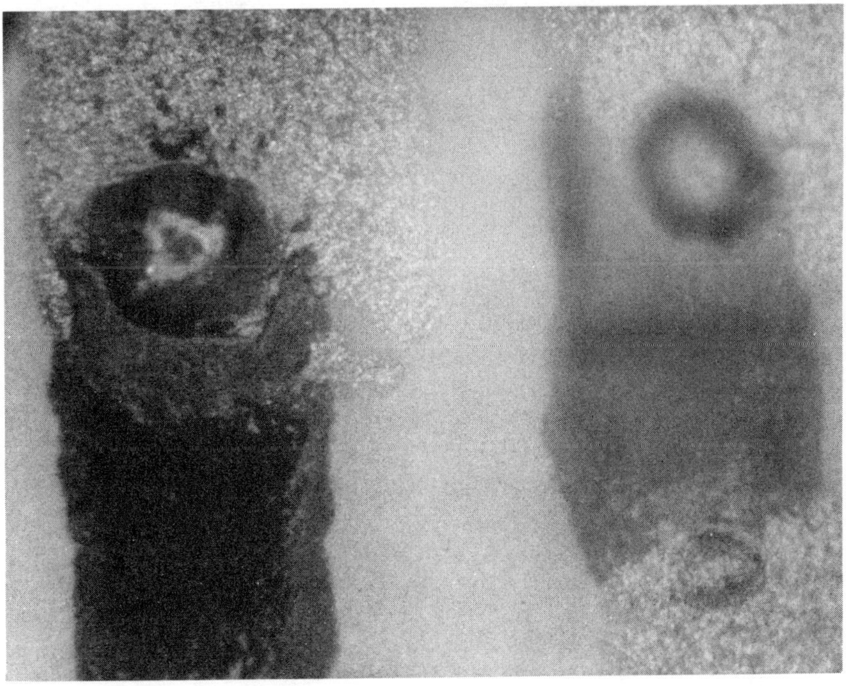

Fig. 6-2. Close-up of three terminations of incompatible resistors and conductors. Note how the resistor material forms a bubble at the resistor-conductor interface after firing.

variables. A particular paste bought at one time may not be the same as the identical type paste bought six months later. The actual amount of variation is hard to predict. Some manufacturers will specify a batch variation of perhaps 10% to 20%. So before even taking the top off the jar of paste, the resistors could be off 15% in value from the last resistor run. Luckily, this problem is seeing constant improvement, as paste manufacturers prepare pastes to tighter tolerances, and paste manufacturing techniques improve.

The next paste variation to be considered is the particular system chosen. There are several in common use.

The traditional basis for thick film resistors has been the palladium-silver system. Palladium-silver has accumulated the most data and is the most widespread. It is reasonably inexpensive, therefore it is ideally suited to general purpose work. The old adage that you get what you pay for is particularly well suited to paste purchasing. Palladium-silver, a relatively cheap paste, can be very difficult to work with, trying to obtain close tolerances. Palladium-silver resistors are very sensitive to firing profile and firing conditions, such as atmospheric effects. The firing of the resistors depends, in part, on the oxidation of the metallic palladium to form palladium oxide. The chemical reaction involved is basically a very touchy one that never reaches a stable state. After firing, if the palladium-silver resistors are ever exposed to a reducing atmosphere, such as hydrogen, the palladium oxide will start to be chemically reduced back to palladium again, and the resistor value will change. This sensitivity to the atmosphere means that palladium-silver resistors always have to be covered with a glaze coat to prevent any atmospheric interaction.

A resistor glaze is a low-melting-temperature, high-glass-content paste that is screened, dipped, or sprayed onto the substrate after resistor firing. The circuit is then fired again at low temperature (not high enough to affect the resistor values) to form a glass seal coat. Another difficulty with the palladium-silver system is involved with abrasive trimming. The resistors are generally glazed after resistor firing. Abrasive trimming will abrade away the glaze coat as well as the resistor material, so the glaze will have to be reapplied to the trim cuts to ensure that the resistors are all sealed again. This just occurs with abrasive-type trimming; pulse discharges will not remove the glaze coat, and laser trimming will seal the cut over again.

The firing furnace time-temperature relationship for firing palladium-silver resistors is critical. A change of only a few degrees at peak firing temperature can make quite drastic changes in the sheet resistance values, especially in the high resistance ranges.

Beyond the palladium-silver series of pastes, it is difficult to make generalizations. The pastes are mostly proprietary. Some of the effects are not exactly understood, and the ones that are, the manufacturers naturally wish to keep to themselves, as this is their way of making a living. The general rule of thumb stated above does hold very true for resistor pastes; you get just what you pay for. The iridium, platinum, and ruthenium systems are more expensive than the palladium-silver pastes, but the results can be far better and easier to work with. Since the more expensive pastes usually have tighter lot tolerances to start with, they can be screened and fired to closer tolerances and the fired tolerance spread is not so wide. The more expensive systems are relatively insensitive to firing profile and conditions, and for most applications do not need a protective overglaze coat. This is not to say that the firing conditions are not important for these pastes, but merely that it is not as critical as for palladium-silver resistors.

At first glance, it may appear that the cheaper pastes would be more desirable in terms of final costs, but this is deceiving. The actual cost of the pastes used, compared to the actual cost of the finished circuit, is very small. Cost could run, for example, 0.1 or 0.02 of a cent per resistor, depending on quantity and complexity. If the paste is changed to one costing three times as much per ounce, the cost per resistor is only up to 0.3 or 0.06 of a cent per resistor. This is still trifling when the transistors may cost 10 cents to a dollar each, and the finished package costs 10 dollars. It may be far more desirable to use the more expensive paste and get better printed tolerances, closer TCR, and eliminate the extra step required for overglazing. Of course this may not be true in all cases. A very low-cost, high-turnover, loose-tolerance circuit may require the cheaper material. This is one of the design and process trade-offs to be considered.

One other general type of resistor paste is the thallium-based system. Thallium oxide pastes are in use and good results have been reported. One reason why the thallium pastes may not be in as widespread use as some of the others is that thallium is toxic. It is a cumulative poison, which means that it is stored in the body. Although it is certainly not as dangerous as many of the chemicals used in the hybrid industry and it can be handled with complete safety, there seems to be a psychological block against handling it and other toxic substances.

Dielectrics

The final screenable paste to be considered is the dielectrics. Dielectric material is used to form crossovers and low-value screened capacitors. A thick film screened capacitor is formed by three screen-printing steps. First a conductor is laid down to form one plate of the capacitor. A dielectric layer is screened on top of this, and then another conductor, to form the other plate of the capacitor, as shown in Fig. 6-3. Multiple screenings and layers can be done. Layers of conductors are alternated with dielectric between them. The conductors are connected alternately to form the plates of the capacitor. One screening step is required for each layer.

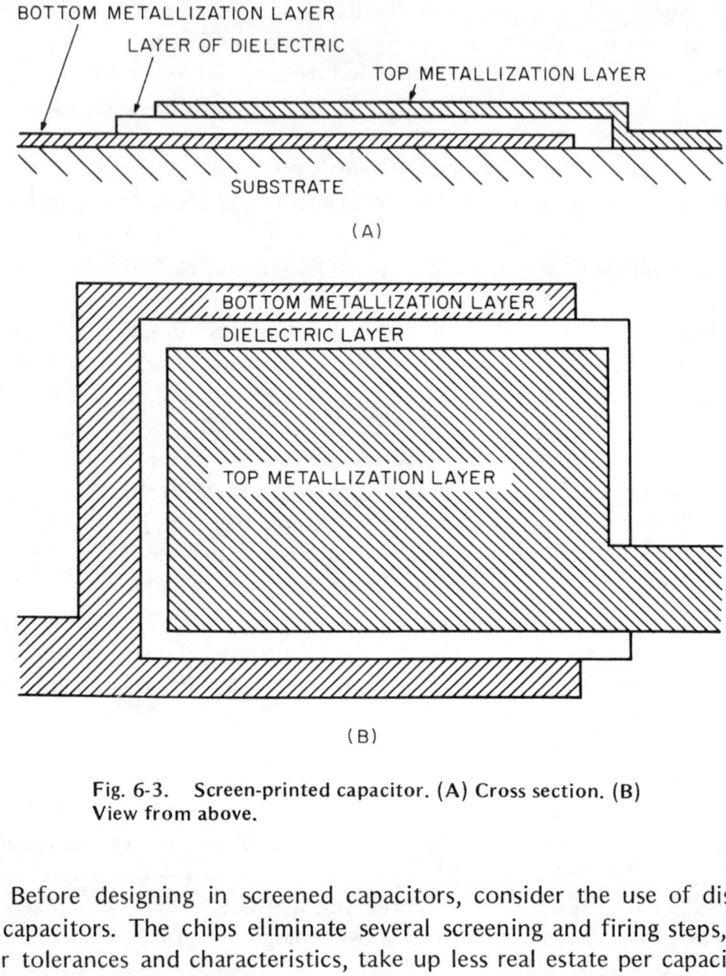

Fig. 6-3. Screen-printed capacitor. (A) Cross section. (B) View from above.

Before designing in screened capacitors, consider the use of discrete chip capacitors. The chips eliminate several screening and firing steps, have better tolerances and characteristics, take up less real estate per capacitance value and are readily available. However, the uses of screened-on capacitors should not be overlooked.

The dielectric pastes used are high-glass-content pastes that form glass coats similar to resistor overglazes. For screening crossovers and capacitors, it is desirable to screen two coats of dielectric material between the conductors as a precaution against pinholing and shorting of the conductors.

The dielectric constant for screenable glazes is fairly low, on the order of a couple of hundred for capacitor dielectrics and less than ten for crossover glasses. Only low-value capacitors can be screened. Screened capacitors also take up quite a large area on the substrate per capacitance value. Higher values can be obtained by multiple screenings of dielectric and conductors in the form of a sandwich, but a limit will be reached when the conductors have a steeper and steeper path to climb up the sides in order to reach the top layers.

There is one final group of pastes that stands by itself. These are the resin pastes. They are low-cost pastes used for mass-produced, high-economy applications. Resin pastes use a resin binder that cures at low temperatures. The substrate material used for these applications need not be ceramic. These pastes can be printed onto epoxy printed-circuit boards or plastic bases, if the support material will stand the curing temperature of the paste. One word of caution to anyone using copper-clad tracks for resistor terminations: The copper oxidizes very easily and will produce a high resistance film at the resistor-copper interface. If copper tracks are to be used, the copper must be carefully cleaned and, while clean, plated or covered with some other material that does not corrode as readily.

Resin-based pastes are limited by their environmental operating conditions; they cannot be used under extreme conditions. Heat is a severe limiting factor. Both internal dissipation and environmental heat will alter the resistor characteristics, due to the low temperature for curing the paste materials.

7 — The Firing Cycle

The thick film conductors, resistors, and dielectrics have been printed and dried. Now they are ready for firing.

Firing is the final stage in resistor- and conductor-making. It takes place at a high temperature, usually between 700° and 1000°C. Firing the printed patterns burns off the organic binder in the pastes and sinters the resistive or conductive particles into a solid mass to form the desired properties. For instance, in a palladium-silver resistor after screening, the printed pattern consists of particles of resistive material held together by the binder. If the resistance is measured, it will be extremely high. When the material is fired, the binder burns off and the resistive material oxidizes and fuses together with the glassy phase to form a palladium/palladium-oxide resistor. If the material is underfired, not enough particles will oxidize and fuse together and the sheet resistance will come out high. If the material is overfired (see Fig. 7-1), the particles all oxidize and fuse together in one glassy sheet, and the sheet resistance will drop way down. The actual firing reactions are much more complex than this simplified picture and involve a complex chemical reduction and oxidation of the paste material. Substrate interactions also enter into the picture.

Firing sounds quite easy. Simply heat the material until the desired results are obtained, then cool it off. It is not quite so easy to do in practice. This is why thick film furnaces are so elaborate.

Methods of Firing

There are three methods of firing: (1) prefiring or pretermination, (2) cofiring, and (3) postfiring or post-termination.

In prefiring or pretermination, the conductors are screened, dried, and fired first. Then the resistors are screened, dried, and fired. The conductor has to fire at a higher temperature than the resistors to prevent undesired

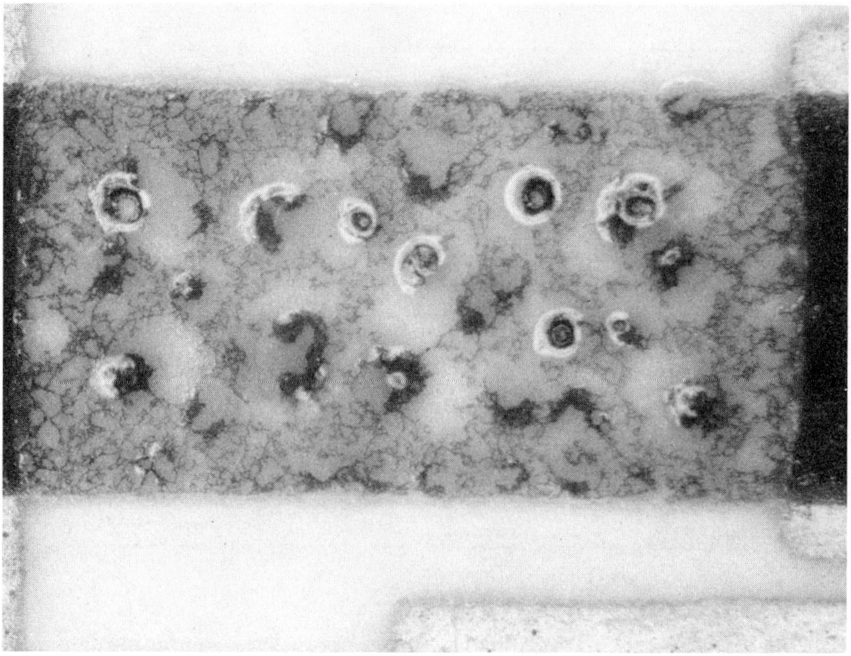

Fig. 7-1. Close-up of an overfired resistor, showing reduction to the glassy state with blistering and cracking in the middle of the resistor.

interactions. When the resistors are fired, the conductors that are already on the substrate are effectively being refired. If the conductor has a lower firing temperature than the resistors, the higher temperature of resistor firing can start off metallurgical reactions at the interface. Prefiring is the generally followed practice.

With cofiring the conductors are screened and dried. Then the resistors are screened and dried. Finally, both the resistors and conductors are fired together. Although this method eliminates one firing cycle, it can only be used if the resistor and conductor pastes have similar firing characteristics. This is not a commonly used method of firing since undesired metallurgical reactions can occur at the conductor-resistor interface.

In postfiring or post-termination, first the resistors and then the conductors are screened, dried, and fired. This method is superior for small resistors. It has the disadvantage of putting the resistors through two firing cycles. If the resistor paste shifts during refiring, then postfiring either cannot be used or allowances for the paste shift must be made during initial design and layout. Postfiring can improve small-resistor printing.

For preterminated resistors, the ideal resistor makes a nice smooth film of the desired geometry, as shown in Fig. 7-2(A) (see also Fig. 6-1). In

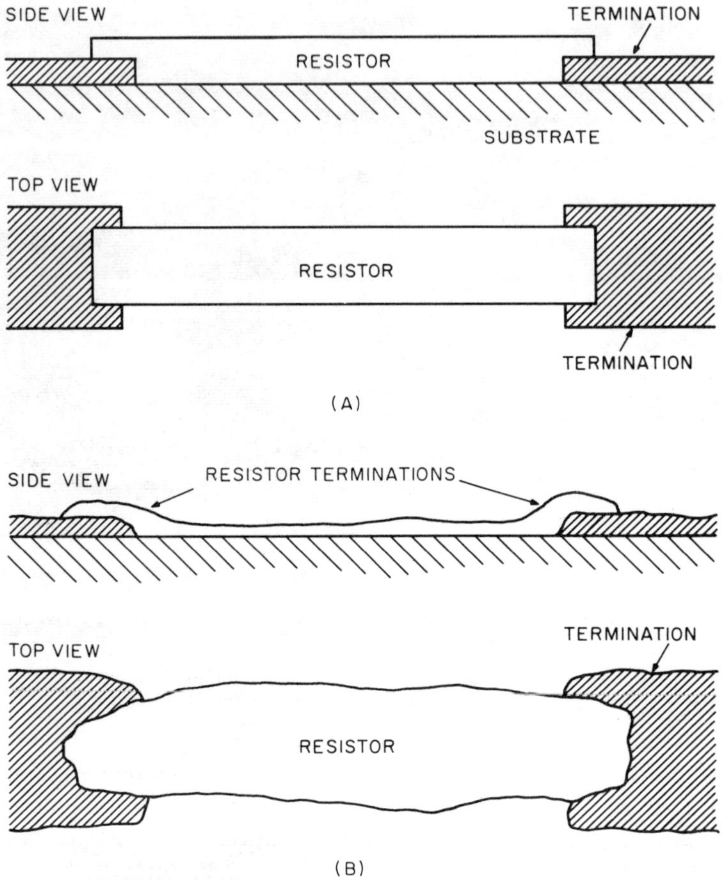

Fig. 7-2. (A) Ideally, when printed, resistors and conductors are geometrical with sharp corners and straight sides. (B) When printed in actual practice, the resistors and conductors do not hold their shapes. The paste tends to settle, creep, and slump.

practice, the paste tends to slump and run, as shown in Fig. 7-2(B). This effect is especially noticeable in small resistors. The resistor will widen where the paste flows down between the conductor terminations. This will change the aspect ratio and thus the resistor value. Slumping will also create problem areas at the end of the conductors, as shown in Fig. 7-3(A). At the top corner of the conductor terminations, the resistor material can become very thin as the material flows down between the conductors. Upon firing, this can result in cracking, blistering, high resistances, and other problems. These effects are usually present to some extent in all resistors, but become more pronounced as smaller geometries are reached.

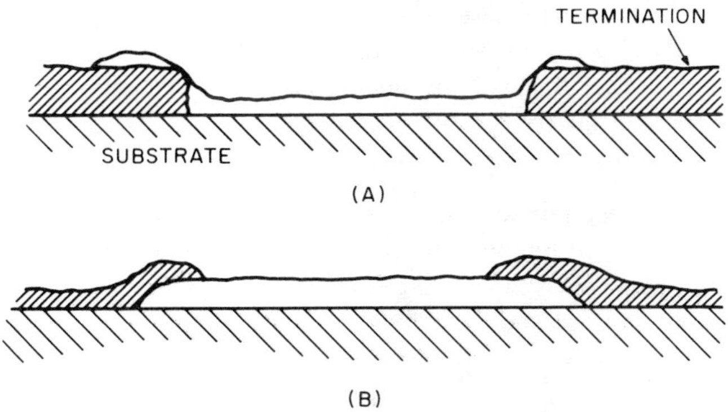

Fig. 7-3. (A) In pretermination the conductor is printed first. (B) In post-termination the resistor is printed first.

When resistors are post-terminated, the conductors cover up the ends of the resistors, as shown in Fig. 7-3(B). The resistor, being screened first, will hold its dimensions much better. If the conductors do slump and widen, it doesn't make as much difference. An added advantage is that the conductor area that overlaps the resistor can be used for bonding or other purposes, instead of being lost under the resistor termination.

The Furnace

Thick film resistors and conductors can be batch-fired in a box kiln. This is very difficult to control. The atmosphere for firing is very important, as interactions take place between the paste undergoing firing and the furnace atmosphere. The heating and cooling rates must be precisely controlled and they must occur at a specified rate and be repeatable. This is not easily performed in a box kiln since a box kiln does not provide good temperature control.

The peak firing temperature is extremely important. It must be closely controlled to at least within one or two degrees. It is not desirable to introduce substrates into a kiln that is in operation at peak temperature. Thermal stresses will occur in the substrate that can fracture or weaken it. The pastes being fired will tend to form a hard fired film right away on the surface. This leaves binders and solvent residues which haven't been burned off underneath this crust. When the underneath layers do burn off, they may crack the film surface or can even explode through it. If the substrates are quickly removed from the peak temperature after firing, they may cool too rapidly and crack or break.

For these reasons, box kilns are not very often used. Instead, continuous belt kilns or furnaces are used. A belt furnace consists of a long heated tube with a moving belt that passes through it. The substrates are placed on the belt and carried through the heated area at a precisely controlled rate. The atmosphere inside the furnace is controlled.

Requirements for firing furnaces are:

1. Ability to keep temperatures constant within ±2°C or better.
2. Repeatability of temperature settings from day to day.
3. Atmosphere control inside the furnace.
4. Constant and repeatable belt speeds.

The heated portion of the furnace consists of a fused quartz muffle surrounded by electric heating elements. Ceramic and metal muffles can and have been used, but are not as satisfactory. The ceramic muffles respond to heat change requirements quite slowly, making it hard to keep the temperature constant under varying load conditions when the substrates start coming through. Metal muffles tend to oxidize with the air passing through, and the high temperatures involved; these oxides flake off the walls of the muffle and contaminate the substrates. New high-temperature materials for metal muffles have cut down on the flaking problems considerably. The furnace temperature is sensed by thermocouples located under the heated zones. The thermocouple output is used to regulate silicon-controlled rectifier systems which in turn control the input power to the heating elements.

The atmosphere inside the furnace must be kept clean. Air currents are used to continuously flush out the muffle. This is to ensure that the binders and remaining solvents in the paste, burned off during the initial temperature rise, are forced out of the furnace and cannot travel down with the substrates to the sintering zone where they might chemically interfere with the firing reactions. Burned-off materials must be blown out of the muffle and clean air drawn in to replace them. There are two methods commonly used for doing this: (1) using compressed air as in Fig. 7-4(A) and (2) tilting the muffle as in Fig. 7-4(B).

The first method involves the use of dry, clean air from a compressed air storage tank. The air is introduced in the furnace at one end and flows on out of the other end. The muffle may include a series of baffles to ensure that the air flows through the way it should. The air moves in a counterflow; it is introduced at the exit end of the furnace and travels back against the substrate movement. It comes out at the entry end of the furnace, carrying with it any contaminants and burned off substances.

Tilting the furnace a few degrees from the horizontal creates the same effect. The exit end is lower than the entry end, so the substrates are actually travelling slightly downhill. Hot air moves upward by convection, pulling out all the contaminants at the top and drawing in clean air at the bottom or exit end at the same time.

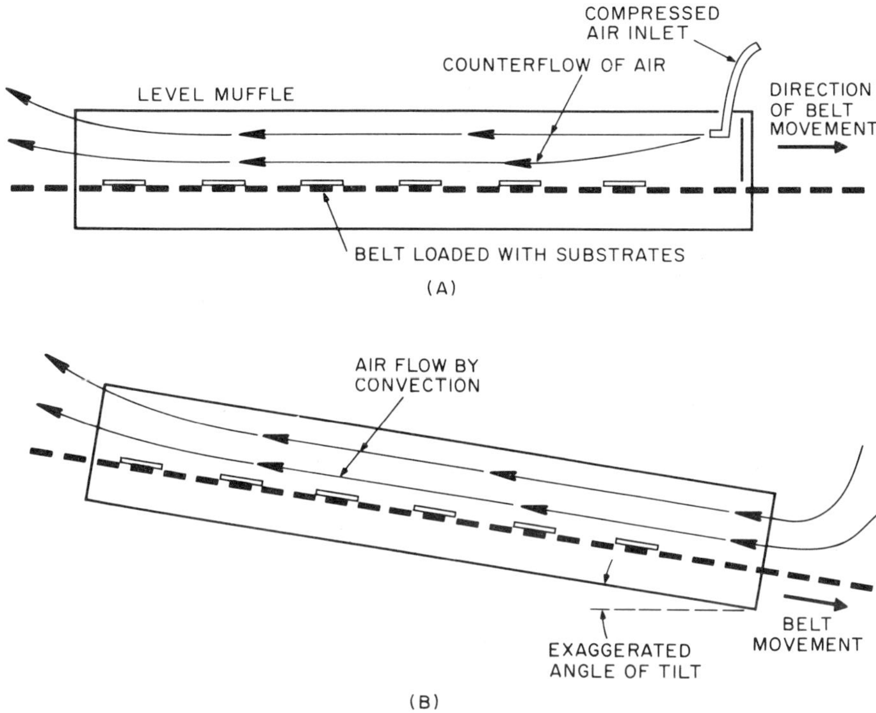

Fig. 7-4. Furnace airflow schemes. (A) Forced airflow using compressed air. (B) Muffle is tilted to achieve airflow by convection.

Most pastes need an oxidizing atmosphere for the proper firing reactions to take place. This is the reason for all the concern over cleaning the air, rather than just pumping nitrogen gas through the muffle. Some techniques do use other gases. Again, the recommendations of the manufacturer should be used.

The important feature of furnace atmosphere control is to make sure that the air going into the furnace is clean. This may sound ridiculous at first, but it is one of those little things that can cause a lot of trouble. When using air from a compressor, careful filtering is necessary to remove all the oil and water vapor. These are present in compressed air, due to the nature of a compressor and its action. If the oil is not removed, it will enter the furnace and be burned. This will form various hydrocarbons that will interfere with the firing process and can produce some very odd results. Other vapors present in the furnace during firing can produce miscellaneous results. Solvent vapors, such as trichloroethylene, if present in a furnace firing a palladium-silver system, will chemically reduce the resistors and will change the sheet resistance values quite drastically.

The exact effect of the presence of water vapor in the firing process is not fully understood. Changing the vapor content does change the fired results somewhat with most pastes. The general opinion seems to be that the humidity is not really so important, but large changes are. It is better that the humidity is kept constant and does not change from day to day. About 50% relative humidity would be a good figure. Changes in relative humidity with changing weather conditions have been known to cause large changes in fired results, causing some processors quite serious problems. This is a consideration when using a tilted muffle approach and using the room air to flush it out. The room temperature and humidity should be kept constant to help achieve consistent results. There must be no chemical vapors present in the room that could be pulled into the muffle during firing.

The furnace belt speed must be closely controlled. Electric speed controls are available that keep the speed constant and repeatable by silicon-controlled rectifier circuits, feedback loops, and tachometer controls.

The Furnace Profile

The heart of the thick film firing process is summed up in the words "furnace profile." The profile refers to a curve that shows the temperature changes that are present as the substrate moves through the furnace. Figure 7-5 shows the zones and profile that may be encountered in a small furnace. Because the substrates are moving on a speed-controlled belt, the profile comes down to the time-temperature relationship that the substrates undergo. It is critical. Both overfiring and underfiring produce undesirable results.

Because the profiling is so important, the furnace is divided into several sections or zones, each with its own temperature-controlling circuitry and heating elements. Typically a small furnace will have four or six controlled zones, but may have as many as twelve. The more zones there are, the more accurate is the profile control. The controller for each zone is set to a temperature that produces the overall required temperature gradient. The overall effect of all the zones is a nice smooth profile.

The actual temperature on the firing belt surface is not what is read out on the controllers. The control thermocouples are not located on the belt where the substrates are, but are usually underneath the hearth somewhere. The effect of sending substrates through on the belt and loading down the furnace, and of blowing fresh air over the belt, is to lower the actual temperature on the belt. This can be as much as 20°-30°C. For accurate and repeatable firing, the temperature must be known right at the substrate. Therefore, some additional means of checking the furnace temperature is needed. The way this is done is by making up a dummy substrate and attaching a thermocouple to it. The substrate is sent through the furnace with extension wires attached, and the output of the thermocouple is fed into a

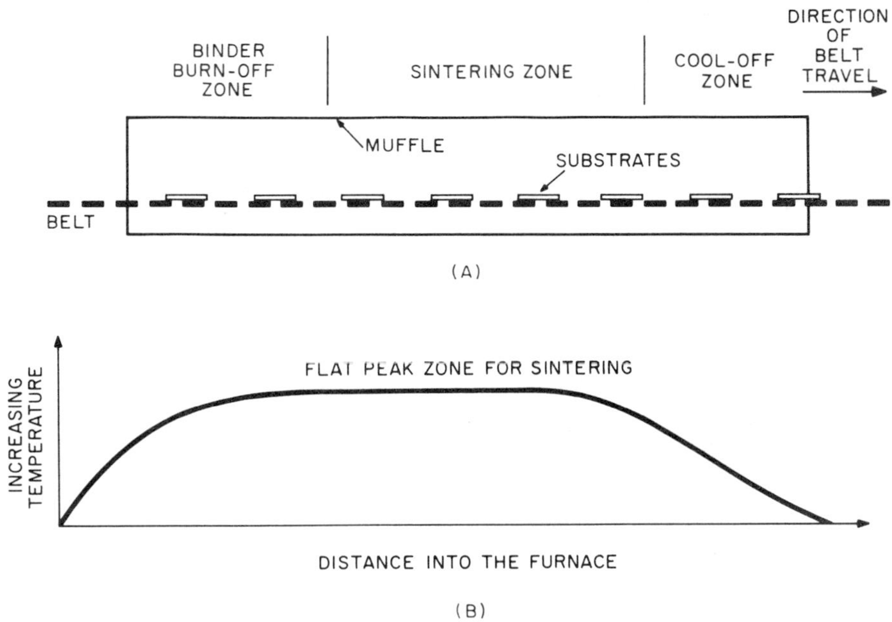

Fig. 7-5. (A) Furnace zones. (B) Typical furnace profile.

chart recorder. This will give a plot of the profile as the substrate and thermocouple go through the furnace. The effect of the substrate is to simulate loading of the furnace. The temperature change between unloaded and loaded conditions can be appreciable. In some cases it may be necessary to load the belt fully, just as though actual firing were taking place, and replace one of the substrates with the thermocouple. Often the substrates are loaded into graphite or quartz boats when they are fired. The mass of the boat and its thermal response will affect the profile. The use of boats facilitates handling of the substrates.

For even more precise measurements, two or three thermocouples (or even more on a very wide belt) are spread across the belt, to record the temperature variations across the belt.

Chromel-alumel thermocouples (type K) are generally used. The extension wire insulation has to be a very high temperature type. Temperatures of up to 1000°C are being measured, and most insulation materials will ignite at these temperatures. One of the common solutions is to insulate the wires with a combination of ceramic tubing and/or ceramic fish-spine beads. This will protect the wire, yet will allow it to flex enough to be handled easily, and be coiled up when not in use. The thermocouple wire should be in a coil no less than 2 feet in diameter. If it is smaller, the wire will be stressed unduly and kinks may be left in it permanently.

The chart recorder will have to be calibrated in terms of the type of thermocouple junction used. Most chart recorders have built-in circuitry to simulate the thermocouple reference junction.

Each paste will have a different profile and peak firing temperature. The individual manufacturer's recommendations will have to be followed to come out with the calibrated values. Sometimes, however, quite satisfactory results or even better results can be obtained by departing from the recommended procedures. Whether they are better or not will depend on the particular process and the desired results. There is usually a trade-off involved. Some characteristics may be improved at the expense of others. The usual trade-offs involved are firing temperature, temperature coefficient of resistance (TCR), stability, and sheet resistance. For example, by increasing the firing temperature with some types of paste, the resistor stability and predictability are markedly improved, but the sheet resistance is decreased and the TCR changes. This is something that has to be worked out experimentally, depending on the paste and processing conditions. The effects of overfiring a resistor are shown in Fig. 7-1.

Summing up the important elements of thick film firing to remember:

1. Buy a furnace that has accurate temperature control, precise belt speeds, and repeatability of both.
2. Use a thermocouple or series of thermocouples to set the profile of the furnace accurately.
3. Make test runs with dummy substrates to determine the effects of belt loading on the temperature.
4. Make sure the air, or other atmosphere used, inside the furnace is clean and dry.
5. Read the paste manufacturer's specification and application sheets carefully. If different results are desired by changing the firing process, this is fine, but it is important to start from the same place as the manufacturer.
6. Have the furnace in a controlled atmosphere to minimize humidity and temperature changes.

8 — Resistor Trimming

Once the conductors and resistors have been deposited and fired on the substrate, they must be tested to see if they are the required values. Electrical inspection reveals that the tight tolerance resistors that were drawn, photographed, printed, and fired with such close controls, have actually ended up with a tolerance spread of ±15% of the desired values. What happened? Unfortunately, this is normal. The tiny variations that creep into the process, and the tolerances of materials and equipment in the various steps leading to this one, make these variations unavoidable. With very close control and the utmost accuracy at each step, a fired variation of ±10% on the resistors is obtainable. Under lab conditions, closer tolerances can be achieved, but for production, a variation of 10 to 20% is to be expected, and can become even greater depending on the materials used and the degree of control over the process. So, what now? How can the 1% tolerance needed for a particular circuit be obtained? The solution is to trim or selectively alter the value of the resistors. The four generally accepted methods of trimming thick film resistors are: by physical abrasion, air abrasion, electric pulse, and laser.

Physical Abrasive Trimming

The scheme behind most of the thick film trimming methods is to remove part of the resistor material, thus raising the resistance. Resistor values are determined by the sheet resistance of the paste used and the number of squares in the resistor. Figure 8-1 shows the effect of making a resistor thinner, and then making it thinner again. If a resistor is made thinner by cutting it in half lengthwise, the size of each of the squares remaining is made smaller. The length of the resistor remains constant, but the number of squares has been increased because of the width change. The sheet resistance has also remained constant, but because the number of squares has increased,

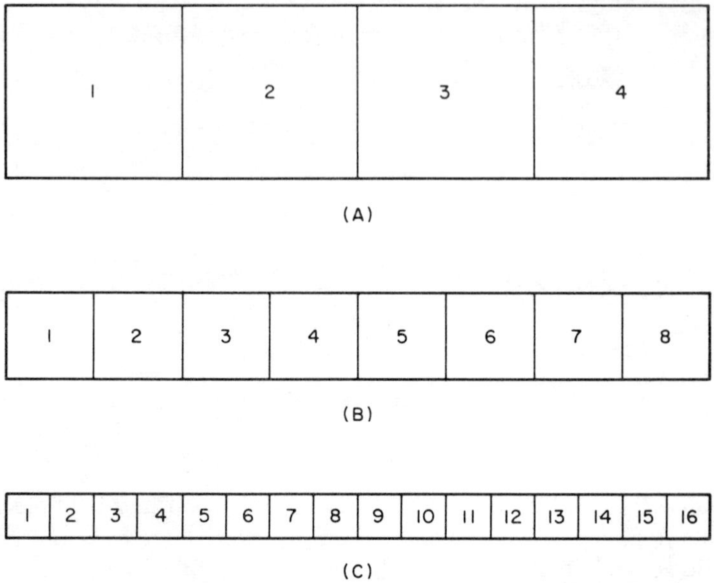

Fig. 8-1. Change of resistance with resistor width change. (A) Typical resistor with a length of four squares. (B) If resistor width is halved, the number of squares doubles. (C) If the resistor width is halved again, the number of squares again doubles. This resistor is one-fourth the width of the first resistor with four times the number of squares.

the resistor value has been increased accordingly. The resistor under consideration has been cut in half, so the number of squares and the actual resistance value has doubled. This is a linear relationship. If the resistor is cut to one quarter of the original width, the overall resistance goes up four times. This is the basis for abrasive trimming—if part of the resistor is selectively cut away, the resistor will increase in value. Figure 8-2 shows the effect of cutting a notch in a resistor. How the actual removal of part of the resistor is done can be as simple or as complicated as desired.

The crudest method of resistor trimming is to use a pencil eraser. This method is only useful on very soft pastes and is extremely inaccurate. There is no way of knowing when the resistor has reached the desired value.

One method that has been used is to chip away part of the resistor with a vibrating point, either a steel needle or a diamond point like a phonograph needle. The point is vibrated at either a low mechanical frequency or an ultrasonic rate. The vibration is used to physically chip away part of the resistor. In its more sophisticated forms, this method is used with electronic bridge equipment to measure the resistance and control the stopping point of the needle.

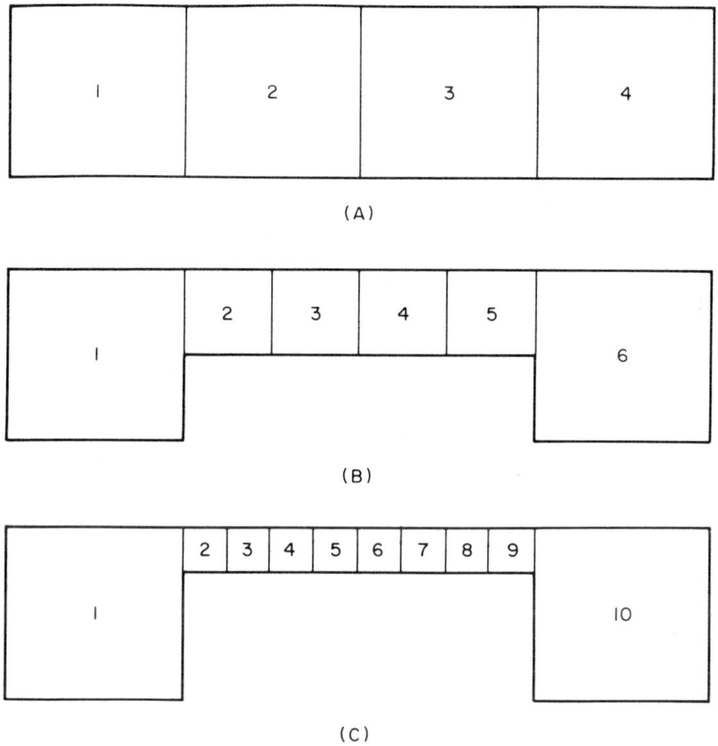

Fig. 8-2. Notch-trimming of resistors. (A) Resistor that is four squares long. (B) If a notch is trimmed out of the middle, the number of squares goes up. The new resistor has six squares instead of four, thus having a value one-and-one-half times that of the first resistor. (C) If the notch is made deeper into the resistor, the number of squares goes up again. This new resistor now has ten squares. The overall resistance value has gone up two-and-a-half times from the original value.

There are several disadvantages to this method: (1) high wear rate of the tool, due to the hardness of the substrate and the film, (2) difficulty of mechanical control, and (3) a jagged cut is left at the resistor edge, raising the resistor noise.

Air Abrasive Trimming

The most sophisticated of the abrasive methods is the air abrasive method. This is actually an extension of the abrasive methods mentioned above. Figure 8-3 shows a representation of an air abrasive trimmer. The abrasive cutting is done by fine gritty particles that are blown by a high-

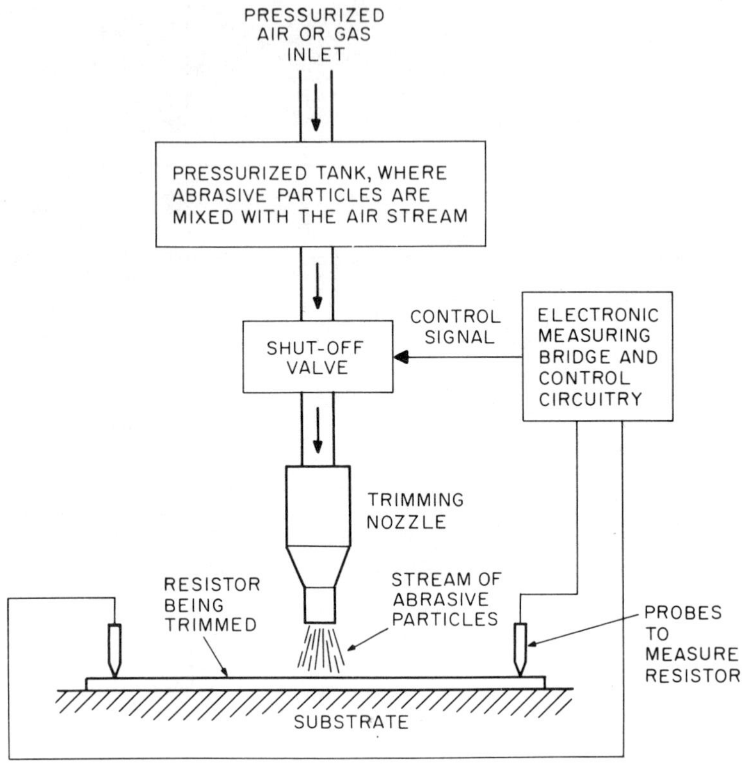

Fig. 8-3. Typical air abrasive trimming station.

pressure air stream onto the resistor to be abraded. This creates a scouring action similar to the effects of sand blown against a house window or car body on a windy day. The jet stream is controlled by a fine orifice that concentrates the abrasive action into a very fine cutting path. The particles used are fine aluminum oxide, silicon carbide, or sodium bicarbonate powders, with a particle size between 10 and 50 microns, depending on the work that is being done. A variety of jet orifices to provide different cut widths and shapes are available, again depending on the geometry and coarseness of the work being done. The air supply for the abrasive generator has to be completely clean and dry. The air is used to force the abrasive powder through the very fine nozzle. Any moisture or oil in the air will cause the powder to clog the nozzle very rapidly. The same goes for the powder itself since any contamination or moisture in it will also cause clogging. One way around the air contamination problem is to use bottled air or nitrogen. For small-quantity work, this will last quite a long time and ensure a clean, dry propellant supply. Moisture can be removed from the powder by baking it at a low temperature for several hours.

Fig. 8-4. Close-up of printed and fired resistor with adjustment notch trimmed out.

The air abrasive system is the most widespread method used for thick film resistor trimming. It is reliable, relatively cheap, well proven, and accuracies of less than ½% can be obtained. The size of the resistor being trimmed is a factor in the accuracy. It is difficult to accurately and repeatably trim a 10 x 20 mil resistor. Narrowest cut widths are on the order of 7 to 10 mils. Wider cuts are obtainable by increasing the size of the jet orifice. An air-abrasive-trimmed resistor is shown in Fig. 8-4.

Air abrasive trimmers have gone through various stages of sophistication because of their widespread use. Complete systems are available to automatically check the resistors, trim them, and mechanically handle the loading and unloading of the substrates.

A typical trimming cycle includes the following steps:

1. The substrate is loaded onto the trimming platform.
2. Probes come down and measure the value of the resistor to see if it is over or under the desired value.
3. If it is under, the airstream generator will turn on and the abrasive orifice will start to move across the resistor, cutting as it goes. If the resistor is over the desired value, no trimming takes place since the resistor cannot be trimmed down in value, only up.

4. The resistance bridge constantly measures the changing resistor value, and when the preset, desired value is reached, the bridge automatically turns off the abrasive generator and recycles the equipment to the start position.
5. The trimmed substrate is unloaded and another one to be trimmed is loaded onto the platform.

The one big disadvantage of the air abrasive system is the abrasive dust itself. A vacuum system is commonly used to continuously collect the dust and particles of abraded resistor as they bounce off the substrate, although there is still an overspray that escapes the pull of the vacuum and covers everything near and far with a fine gray dust. In the case of beryllia substrates there is an additional problem, because the dust itself is toxic and additional precautions have to be taken to protect the operator from this. Another disadvantage is that active devices such as transistor chips have to be added after the trimming operation, since overspray of the abrasive dust will abrade them away too and can either change their characteristics or ruin them completely. Thus it is difficult to actively trim a circuit while it is operating, to get a certain voltage or current or other active parameter.

All the mechanical abrasive methods actually cut away a portion of the resistor, raising the resistor value. The resistor value cannot be lowered by these mechanical means. Abrading away part of the resistor leaves an exposed, cut edge. In the case of palladium-silver resistors or other resistors susceptible to environmental effects, this edge will have to be sealed over again with a glaze coat after trimming. Since a glaze coat generally has to be put on before trimming for greater accuracy, two or more glazing steps are necessary.

The fact that abrasive resistor systems can only trim a resistor up in value is in itself a drawback. The resistor has to be designed to less than its final value to ensure that the resistor can be trimmed. As mentioned earlier on, if the resistor is laid out at its actual value, the fired value may come out too high and be outside the trimmable or usable range.

Electric and Electrochemical Trimming

There are several electric and chemical methods. One is to pass an electric current through the resistor to fuse some of the molecules together. Usually a pulsed voltage source is used for this. Equipment is available to automatically measure the resistor and control the trimming source, as in the abrasive method. Two advantages of this electric method are that it is clean and the resistor size is not a problem. The major disadvantage is that it is suitable only for certain types of resistor paste. Moreover, the resistors cannot be operated above the trim voltage or the resistance-changing action will continue to take place. There is some question about the long-term stability of resistors trimmed this way since they appear to be more susceptible to drift than resistors trimmed by the air abrasive method. This type of trimming has, up to this point, been used primarily for trimming thin film resistors. Very little work has been done with thick films and a prospective user would have to evaluate the equipment in terms of the paste and system he is using. This will change as more results are published.

Electrochemical trimming, again used primarily with thin film circuits, uses an electrolysis action from applied chemicals to selectively reduce or oxidize the metallic elements in the resistor films, thus raising or lowering the resistance value.

Laser Trimming

The final method, and the one that seems to be gaining the most popularity, is laser trimming. Laser trimming uses the heat generated by a concentrated laser beam to vaporize part of the resistor. As with the other methods, probes and electric bridges are used to measure the resistor and control the laser cut. Various mechanical methods are used to move the substrate under the beam. Alternately, the laser head can be moved across the substrate while the substrate is held in a fixed position.

One of the benefits of laser trimming is that resistor values can be either increased or decreased. Laser trimmers operate by two methods, depending on the amount of heat supplied to the resistor. If low heat is used, most of the resistor pastes can be annealed, which lowers the resistor value. If high heat is used, the laser beam will vaporize the resistor material and remove it just like the air abrasive trimmer, increasing the resistor value.

Lasers also have the advantage of being a clean trimming method. There is a small amount of debris that results from the vaporized material. This can be handled easily by having a small stream of air or other gas flowing across the substrate, to prevent the material from being redeposited on the substrate or clouding up the laser lens. The amount of material is very small and is not a problem. Compared to the air abrasive method, the laser is totally clean. Another advantage of the laser is that it can produce very fine cut widths, allowing trimming of very small geometries. As the cut is made, the heat generated is enough to reseal the glaze coat behind the beam, eliminating the need for reglazing. Because of the initial equipment investment, laser trimming is relatively expensive compared to some of the other methods, but high speed, accuracy, and cleanliness provide superior capabilities.

Unlike some of the other kinds of trimmers, laser trimmers can be used to actively trim a circuit. There is not the risk of the overspray affecting other components as there is with the air abrasive system. The circuit can be built up and then the laser is used to trim a resistor or resistors to provide a particular function, such as a current or voltage gain, while the circuit is operating. This is opposed to the methods where the resistors are trimmed to a specified value before the active devices are assembled onto the substrate. Resistor tolerances of less than 0.1% can be obtained using a laser.

There are two types of laser in general use for thick film resistor trimming: the YAG and the CO_2.

Since there are plenty of good texts and articles that deal with laser operation, a detailed explanation of lasers will not be given here. However, a short explanation of lasing may be of interest and will help in understanding the equipment. Lasers operate on the principle of energy on an atomic scale being converted into optical energy or light. Electrons are "pumped up" away from the nucleus either by optical or electrical means to give them large amounts of energy. Since the electrons end up in a basically unstable, high-energy state, they tend to fall back "down" to their original stable, lower energy states. As they make this transition, they release their excess energy as photons, or what we see as light. The amount of energy that is given up will correspond to a particular photon wavelength or color of light. The electrons are pumped up the same distance each time, releasing the same amount of energy each time, and giving rise to the same optical wavelength each time. The laser beam is monochromatic—it has a single light frequency or color.

This energy-change cycle takes place in a specially formed optical cavity, as is shown in Fig. 8-5. It is constructed at a multiple of the photon wavelength and has reflecting, parallel, silvered faces at each end of the cavity. The cavity acts as an optical resonator, so standing waves will build up at the photon wavelength. One end of the cavity is made to be totally

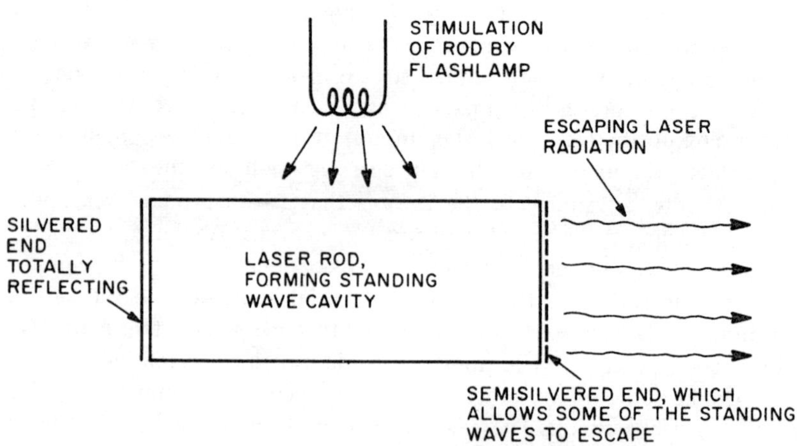

Fig. 8-5. Solid rod laser operation.

reflecting, but the other end is only partially reflecting, allowing the light to escape from it in a narrow, high-energy beam. This beam is then shaped and directed by suitable optics, to perform useful work.

In the YAG laser, the active material is a neodymium-doped yttrium-aluminum-garnet rod. It is pumped by use of a high-intensity flash lamp. The output wavelength is 1.06 microns. This is in the optical region and can cause extensive eye damage. If observed without suitable protection, the high intensity beam will penetrate to the back of the eye and cause irreversible damage to the retina, similar to that caused by looking directly at the sun. Partial or total loss of vision will occur. YAG lasers must always be used with suitable protective equipment, either to optically shield the complete operating setup or through the use of individual protective goggles. An alternative scheme uses closed circuit television to monitor the working area and remote controls to operate the equipment.

YAG lasers have several advantages over the CO_2 lasers. Since the beam produced by the YAG laser is in the optical region, ordinary glass lenses can be used to focus it. For the same reason, YAG lasers can be used to trim through glass overglazes on resistors and capacitors. The YAG laser wavelength is much smaller than the CO_2 wavelength, so the YAG can be used to produce finer cut widths than the CO_2. YAG lasers will produce cut widths of around 1 mil and can be used down to about ½ mil. The disadvantages of the YAG laser are the safety requirements, a lower optical conversion efficiency than CO_2 (meaning higher input power levels and often cooling requirements), and YAG at this time is considerably more expensive. If operated correctly, the YAG laser rod will last almost indefinitely.

In a CO_2 laser the active material is a gas mixture of nitrogen, helium, and carbon dioxide. The gas atoms are raised to the excited state by passing electrical energy directly through the mixture. CO_2 lasers are available as continuously fed gas lasers or with a fixed amount of the gases already contained in a sealed tube. In either case, the gas has to be replaced to offset losses—it is replenished in a continuous flow or, in the case of the sealed tube, the entire tube assembly is replaced after a certain number of hours of operation. The sealed tube can last up to about a year depending on operating conditions.

The CO_2 laser operates at a wavelength of 10.6 microns in the far infrared region. Because of the longer wavelength, the cuts will be coarser than a YAG laser's, usually about 5 mils. The big disadvantage of CO_2 lasers is that because of the operation in the infrared region, glass is optically opaque to the laser beam. Various types of crystals are used for beam shaping and focussing. This means that resistors with an overglaze coat cannot be trimmed with CO_2 lasers and YAG must be used instead. However, this opacity proves to be an advantage from the safety point of view, since ordinary glasses or microscope optics can be used as safety glasses. The eye

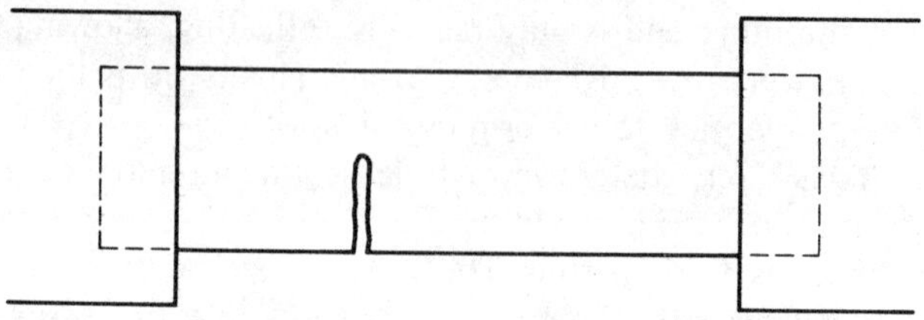

(A)

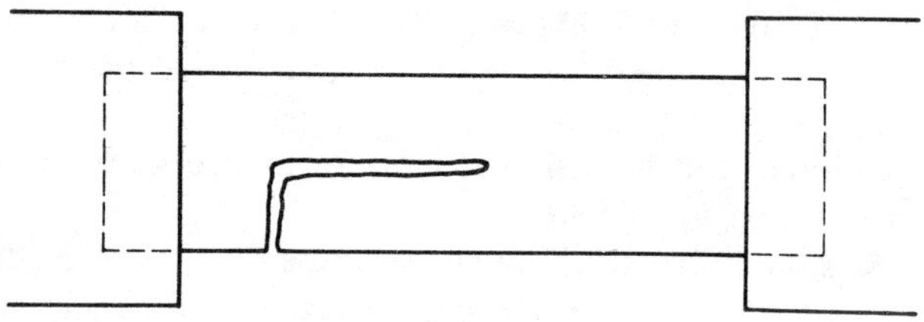

(B)

Fig. 8-6. Types of cuts used with laser trimmers. (A) Plunge cut. (B) L-cut.

itself is theoretically opaque to the CO_2 wavelength, so the chances of eye damage are remote. However it never hurts to be cautious and take all possible precautions.

A third type of laser that is sometimes used is the argon laser. This produces green light at 0.5 micron. The argon laser also requires optical shielding for operator safety. The wavelength is half that of the YAG laser, so correspondingly narrower cut widths can be obtained.

Lasers, for practical thick film trimming use, are generally operated in the Q-switched mode. This means that the laser beam is turned off and on, by electro-optic or mechanical means, several hundred times or more per second. Trimming is sometimes done in the continuous (CW) mode, but generally this generates too much heat buildup on the substrate. This excess heat leads to cracking of the substrate and the resistor film or an undesirable alteration of the resistor characteristics. These resistor changes can be just temporary, which will lead to faulty operation and shut-off by the monitoring circuitry. In the Q-switched mode, the substrate has a very short period of time to cool off between laser pulses. The pulses may operate up to peak powers of 80 or 90 kilowatts, but the short-duty cycles of the pulses reduce the average power to only a few watts. In this way the heating problems on the substrate are eliminated and operating hazards are reduced.

The laser trimming method that has gained the widest acceptance is called the L-cut method. This type of cut is actually two cuts in one. The

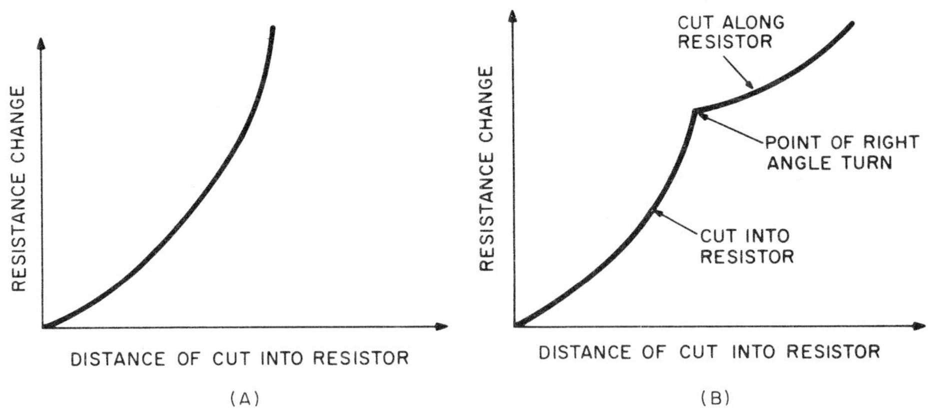

Fig. 8-7. Resistance change versus cut for laser trims. (A) Plunge cut. (B) L-cut.

laser beam starts into the resistor at right angles. After it has gone in a certain predetermined distance, it changes direction by 90 degrees and cuts along the length of the resistor. The L-cut method allows finer control over the trimmed value than the straight-in or "plunge" cut. The two types of cut are diagrammed in Fig. 8-6. The reason for the finer control is seen in Fig. 8-7. The left diagram shows the resistance change versus distance for a plunge cut. The rate of change of resistance increases rapidly, making it difficult to accurately zero in on a preset value. However, if an L-cut is made, the rate of resistance change is decreased as the cut goes along the resistor length. This is shown in the right diagram. The point at which the cut direction changes is precalculated and programmed into the trimmer controls.

There are still some questions about the long-term effects of laser-trimmed resistors. There may be problems with drift and noise with some of the thick film pastes. Some pastes have been shown to be very stable, and others seem to need further work. Other problems that have been reported are the occurrence of fissures, particle clumping, and microcracks, usually produced by localized overheating of the resistor and substrate. As with other aspects of the thick film game, the user will have to evaluate his choice of pastes, substrates, and process in terms of his laser equipment and associated equipment and of the circuit being produced.

Laser trimmers require a large initial investment, but for high-volume, continuous production, the results can be excellent and the actual costs per trim very reasonable.

9 — Active and Passive Components

The thick film hybrid circuit at this point consists of a conductive pattern laid out on the substrate with resistive elements, dielectric films, and crossovers in the appropriate places. To complete the circuit, a variety of other elements may be added. Transistors, monolithic integrated circuits, chip capacitors, chip or subminiature inductors, and chip resistors are the common elements used.

Active Devices

The transistor and integrated circuit elements used in thick film circuits are the same ones that semiconductor manufacturers use in their discrete lines. The manufacture of transistors begins with a thin wafer, some 1 or 2 inches in diameter, of high-purity silicon. Most transistor dice are around 15 to 25 mils square, so several hundred can be made at one time on a single silicon wafer. The transistors on the wafer are sample-tested before separation so the general characteristics of the transistors are known and the percentage of good devices can be calculated. Then the silicon wafer is either scribed with a diamond point and broken apart or the wafer is sawn up with an abrasive slurry. Laser cutting is also under investigation. The result is many single transistors called chips or dice. Each die is mounted on a header, such as the base of a TO-5 can, and wires are attached between the package leads and the active areas. After this the package is sealed up. Each transistor is tested, usually on a computer-controlled automatic machine, and sorted into groups with various characteristics. One group might have MIL-SPEC performance characteristics, whereas another would be adequate for commercial operation. Prime-quality electrical characteristics, beta grouping, noise performance, breakdown voltage, or any other characteristic may be the basis for the grouping. Each group is individually labeled with a JEDEC or manufacturer's number to correspond to certain electrical specifications and package dimensions. So it can be seen that what started out as a single wafer of dice could end up as several different transistor types. By the same token, a single transistor number could have different die types in its packages. This is a factor to be kept in mind when looking for a chip counterpart for a discrete transistor. There may be several chips that will meet the requirement. Also to be considered is the fact that a chip transistor with a 2N number may not be the same as the packaged discrete device. Package effects such as capacitances enter into the picture. Degradation of the transistor characteristics caused by later die bonding will also be a factor to some degree.

To understand all this better, a brief look at monolithic processes may help. Typical transistor-making process steps are outlined in Fig. 9-1. Manufacture starts with a thin wafer of silicon. Impurities are diffused into this basic silicon wafer at high temperatures, to form n-type or p-type material. The bottom of the silicon wafer will become the collector of the transistor, so for an NPN transistor, the impurity diffused into the wafer would be an n-type or donor type of impurity.

Next, a thin layer of silicon dioxide is grown on the surface of the wafer. This oxide layer is selectively etched away by the use of photoresist and an acid etch. The result is a pattern of silicon dioxide and open areas of doped silicon. The silicon dioxide acts as a masking pattern for another diffusion. For the NPN transistor, the next diffusion would produce p-type material in the areas unprotected by the silicon dioxide. This p-type material becomes the base of the transistor. Another layer of silicon dioxide is grown and then etched to give a mask for the final n-type diffusion that forms the emitter. These masking and diffusion steps form an NPN structure for the emitter-base-collector structure of a transistor. For a PNP transistor, the impurities used in diffusion would be the opposite type: p-type instead of n-type and vice versa. The exposed areas on the top after the final diffusion are metallized to form the electrical contacts for the base and the emitter. The metallization is usually aluminum. Aluminum is easily deposited, it has good adhesion to silicon and forms a low-resistance, ohmic contact. This is not the only way of making a transistor and is a simplified version of a very complex process, but the description serves to illustrate some of the steps of active device manufacture.

Because of the construction process, the collector contact is the bottom surface of the silicon chip. The later die-bonding step to attach the transistor chip to the hybrid circuit actually forms the electrical collector contact to the circuit, as well as being the mechanical mounting to the substrate. Usually the chips come with a gold metallization several thousand Angstroms thick, that has been evaporated onto the back or collector surface. This aids in die bonding the chip onto the conductors. As well as being a help, this metallization can also be a hindrance. If the silicon wafer is not properly cleaned or

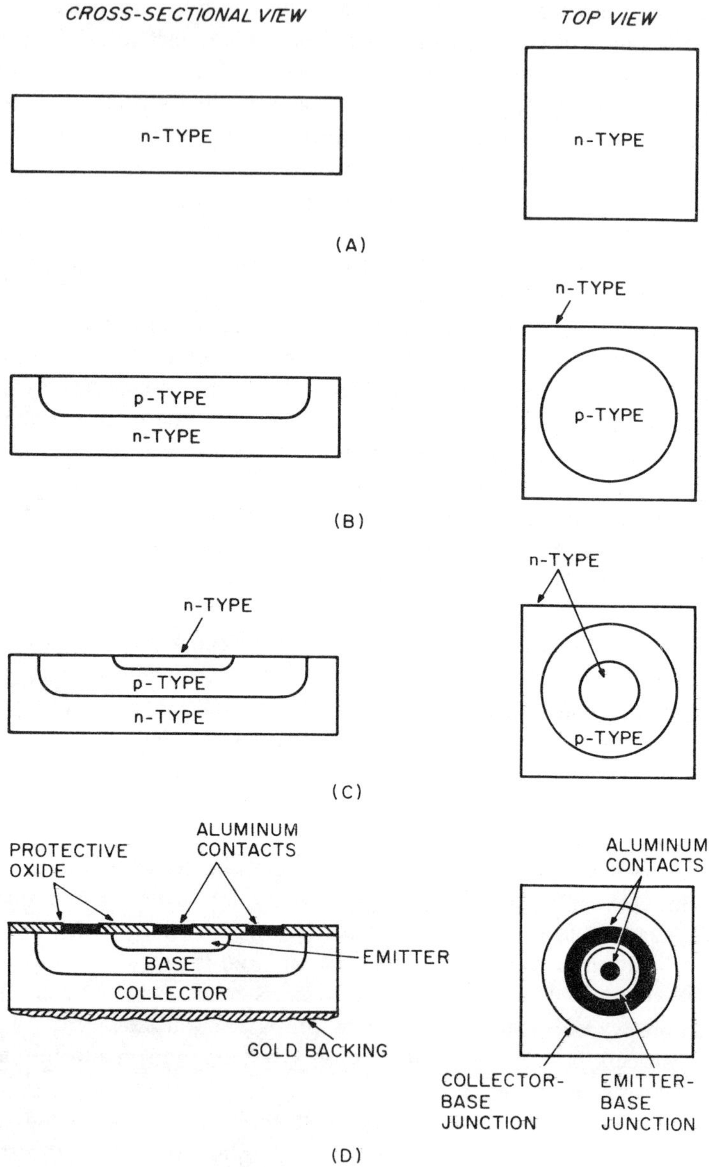

Fig. 9-1. Typical NPN transistor manufacturing steps. (A) Block of doped silicon which will become the collector. (B) Base area is diffused into silicon block. (C) Emitter area is diffused into base material. (D) Protective oxide is grown over the junctions. Metallization is evaporated on to make electrical contact.

if any silicon dioxide is left on the surface before the gold layer is deposited, either the chip will not adhere to the conductor or electrical contact will not be good or may even result in no contact at all. Cases have occurred where all the gold metallization had to be abraded away and the bare silicon bonded to the pads to achieve good results. This aspect is the responsibility of the transistor manufacturer and the integrity of his processes. Normally this is not a problem, but it is good to be aware of the possibility. The top of the chip has its share of problems, as well as the bottom of the chip. For good results the chips have to be passivated or protected from the atmosphere. This is especially true of chips that will go into a nonhermetic package, where moisture and environmental gases could attack the chips and degrade their performance. A film of silicon monoxide or silicon dioxide (glass) is used. If the process is not performed carefully there is a tendency for pinholes to form in the oxide layer which will defeat the purpose of the passivation. Silicon nitride passivation is becoming a more common technique for die protection as it appears to be superior to the silicon oxides.

Semiconductor diodes use basically the same diffusion processes except that the emitter diffusion step does not take place. Only one junction is formed and one contact made on the top surface.

Integrated circuit chips, similar to their discrete versions, are available from most semiconductor manufacturers. Like the transistors they also consist of a block of silicon with bonding pads on the top surface. Impurity diffusions are made with select geometries in various areas of the integrated circuit chip to form various transistors and resistors. Metallization patterns connect the various parts of the circuit as well as form bonding pads for connection to the outside. The monolithic processes for integrated circuit manufacture are very involved and are too complex to go into the details here, so it is sufficient to say that the manufacturer will have to be consulted for interconnects, process, and geometry. The same is true for field-effect transistor chips.

Active devices are nearly always monolithic fabricated devices that are attached after paste firing. Nonetheless, work is being done on both thick and thin film transistors. Thick film field-effect transistors and thin film transistors have been made. At this point, however, both types are merely laboratory devices, produced in experimental quantities only. Their characteristics are generally quite poor compared to monolithically produced devices, and considerably more work will be required before these devices reach production expectations.

Storage and handling of active devices is worth a moment's consideration. Uncased dice are received and used with the metallization exposed. This metallization, usually aluminum, will oxidize quite rapidly when exposed to the air. The oxide films that are formed are quite tough and thick and can make it difficult to die- or wire-bond these devices. In bad cases of oxidation,

Fig. 9-2. Active device carriers. On the left is a compartmented tray, also called a waffle carrier. On the right is a bulk-packed glass vial.

bonding may become impossible. It is therefore desirable to store these devices in an inert atmosphere, such as nitrogen gas, to prevent the oxidation.

There are two common ways that transistor chips come from the semiconductor manufacturer, either bulk-packed in a vial or in compartmented trays. These are both shown in Fig. 9-2. For bulk shipment, up to around 10,000 transistor dice are placed in an ampoule like those used for supplying certain medications and hypodermic dosages in the medical profession. The vial is then backfilled with nitrogen and sealed. The chips inside will be protected by this inert atmosphere until the vial is opened. To open it the neck is snapped off around a prescored line. When the vial is opened, the dice have to be stored in an inert atmosphere. Small storage cabinets are available for this. They are completely sealed, including seals around the doors, so that the cabinet can be filled with nitrogen. A flow rate of only one or two cubic feet per hour of nitrogen is used, so it does not require much gas.

The bulk-packaging scheme is mechanically hard on the individual chips. There is no individual protection so the chips can become scratched, cracked, and chipped quite easily. The motion of the chips grinding against one another will cause a fine silicon dust to appear all over the devices. This has to be cleaned off before the chips are used. To avoid these mechanical problems, small compartmented trays are also used. Each compartment is large enough for one chip. Compartments of different sizes are used for different-sized chips, so a tray may hold from 30 to 600 chips depending on the size. The trays are covered with a thin sheet of antistatic polyethylene and then another plastic tray forms a protective cover. These trays are not hermetically sealed, requiring that they be stored in an inert atmosphere also.

The physical handling of the chips themselves requires special care. It is not very easy to pick up a 20 x 20 mil chip without damaging it. One solution that comes to mind is to handle the devices with some very fine-pointed tweezers, but it is very easy to scratch the die and ruin it with tweezers—it is also not as easy as it may seem to pick up these tiny segments of silicon. About the safest way to handle the dice is with a vacuum probe. This is just a small vacuum head about the size of a pen with a tiny vacuum opening like a hypodermic needle at the end. The vacuum can be turned on or off to pick up or release the chip as desired. Often the actual pickup tip is an inert, soft material such as teflon, to prevent damage to the die.

Passive Devices

There are certain characteristics of some circuit elements that are not readily achievable through screening techniques at this time. Such circuit elements are large-value capacitors, very high- or very low-value resistors, and large-value inductors. The only way to incorporate these in a thick film circuit is to add them later in the form of discrete microdevices. Such components are chip resistors, chip and microcapacitors, and chip and miniature inductors.

Chip resistors are used for two reasons. They conserve space and can provide values not readily obtainable by printing techniques. Chip resistors are produced by three techniques: monolithic diffusion, thin film deposition, and thick film printing.

Monolithic techniques use the properties of the bulk resistivity of silicon by diffusing an impurity with a known geometry into a bar of silicon and depositing metallized pads at each end of the bar to form external connections. The basic technique is the same as for transistor diffusions—photo-resist and silicon dioxide masking, then diffusion of impurities into the bar at high temperature. The range of resistances obtainable is on the order of 100 to 30,000 ohms. It must be remembered that this technique forms a P-N junction, just like a transistor junction, between the resistor (the diffused-in impurity) and the host bar of silicon. This may be undesirable if the bottom of the silicon bar is not isolated electrically from the rest of the circuit. Possible forward biasing of the junction could occur. This type of resistor has several parasitic components associated with it that may interfere with circuit operation. Diffused resistors are not commonly used in thick film circuits.

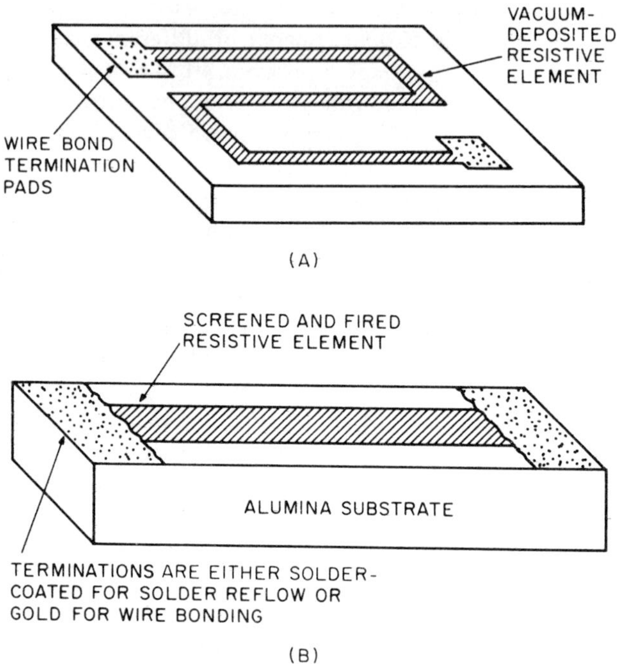

Fig. 9-3. Chip resistors. (A) Thin film. (B) Thick film.

Thin film resistor chips, as in Fig. 9-3(A), are made by vacuum-depositing a thin layer of resistive material onto a 99.5% alumina or a glass-coated substrate. The commonly used resistive materials are nichrome (nickel-chromium alloy) or tantalum nitride. Termination metallizations of aluminum, silver, or gold are added and sometimes the entire chip may be gold-backed. Chip sizes on the order of 50 x 50 mils are common. Thin film resistor chips values from about 10 ohms to 500,000 ohms are available. Resistors are available outside this range, but are more difficult to produce. On the high end, the chip sizes can become quite large and usually more expensive. The chips themselves are attached by eutectic bonding in the case of the gold-backed chips or can be fastened in place with epoxy and wire bonded to form the electrical connections. Chips are available with solder terminations for either conventional soldering or reflow soldering. Reflow soldering is used to form electrical as well as mechanical connection at the same time.

The third type of resistor chip that is used is the thick film resistor chip, as in Fig. 9-3(B). Compared to the thin film chips, the range of values is improved on the high end—up to 1000 megohms can be achieved. These resistors are just the same as a regular screen-printed resistor. A series of resistors is printed onto a pre-scribed substrate. After firing, the substrate is broken apart to form many small resistor chips from about 40 x 40 mils up to 100 x 100 mils and larger, depending on the power dissipation and the resistance required. As with the thin film chips, terminations are available for either chip- and wire-bonding or for solder techniques.

Discrete resistor chips, used wisely, are very useful. They can be used to save substrate area. By proper choice in certain situations, a much higher resistance per unit area can be obtained than by printing directly onto the substrate. This would be advantageous if an entire circuit with high resistor values were to be built in a TO-5 can. The disadvantage of this approach is that wires have to be bonded to the chips to form electrical connections. This increases the possibility of failures by introducing additional bonds. Chip resistors are becoming available in beam-leaded form. This type has two metallic leads projecting from the chip body, which are bonded directly to the conductors to eliminate wire interconnections. Another disadvantage is the ever-present specter of costs. Chip resistors are considerably more expensive per resistor than printing and firing a resistor onto a substrate. For prototyping or for low-volume production, chip resistors may be a cheaper way to go than the entire process of making screens and printing and firing the resistors. In certain situations the advantages of chip resistors can outweigh the apparent cost factor. These chips are very handy when a special resistor value is needed in a circuit. The requirement could be a very close tolerance, or a very high or a very low value of resistance. There may be a case where a single 10-megohm resistor is required in a thick film circuit. Rather than go through a separate screening, firing, and trimming operation for this single resistor, it may be cheaper to add one chip after the rest of the resistors have been fired and trimmed.

As mentioned above, chip resistors can be attached to the circuit by a variety of methods. Since thin film chips and monolithic chips are generally supplied with gold backing, they can be eutectically bonded into the circuit the same as bare semiconductor chips. This means that they can also be bonded with conductive or nonconductive epoxy. Termination pads at the ends of the resistor are connected to the circuit by wire bonding. Chip resistors are also available with beam leads, so beam-lead bonding machines can be used for circuit attachment. Thick film chips are generally either epoxied to the substrate or supplied with metallization that comes over the side of the chip, allowing it to be attached by solder reflow or microsoldering.

The final decision of whether or not to use chip resistors will depend on the particular process, requirements, and volume.

The case for using chip capacitors is far better defined than for chip resistors. The dielectric constant for the capacitor pastes used in thick film screened capacitors is fairly low in comparison to capacitance requirements. Dielectric constants of several hundred are obtainable, resulting in a maximum practical capacitance of around 0.1 microfarad. When high-value capacitors are required, either the capacitor ends up taking over the major area of the substrate or the capacitance is completely unobtainable. The solution is to go to a chip capacitor element. There are three commonly used approaches: monolithic capacitors, multilayer ceramic capacitors, and tantalum slug capacitors.

It is sometimes necessary to provide extremely small capacitances without using screen printing techniques. The type of chip that does this best uses monolithic fabrication techniques to form an oxide dielectric capacitor. A monolithic-oxide type of capacitor chip starts out with a bar of low-resistivity silicon. Using monolithic techniques, a thin layer of silicon dioxide is grown on top of this. Then a layer of aluminum metallization is deposited on top of the oxide. The aluminum metallization and the silicon-base material form the capacitor plates, with the silicon dioxide acting as the capacitor dielectric between them. Capacitors up to several hundred picofarads can be made by this technique.

Another of the monolithic techniques uses transistor-type diffusions to form a P-N junction capacitor. When a P-N junction has a reverse bias applied, the internal space charge region arranges the charge in such a way that the junction acts as if a capacitor were there holding the charge. This junction can either simulate a fixed capacitor or, if the reverse bias voltage is varied, the junction will act like a very small variable capacitor, the actual capacitance depending on the magnitude of the reverse bias. The junction has to be kept in a reverse biased state. If the junction is forward biased, the normal diode rectification action will be observed and the capacitive action is lost.

The next step up in the range of values is the ceramic multilayer type of capacitor. These are made by green ceramic processing techniques. The capacitor consists of alternating layers of dielectric and conductor that are fired to form a sintered mass. The conductor layers are alternately brought out to each end of the capacitor and then shorted at each end with gold or silver metallization to form the two capacitor connections. The commonly used dielectric is barium titanate ($BaTiO_3$). This has a relatively high dielectric constant of up to 6000 or 8000, depending on the processing conditions. Higher and lower dielectric constant materials are available, but usually are not suitable. The lower dielectric constant materials provide less capacitance per unit area leading to lower volumetric efficiency; the higher ones tend to be unstable under changing temperature conditions and may drift with time. Work is also being done with glass-ceramic dielectric layers to produce about the same results.

The finished sandwich (Fig. 9-4) usually has silver end metallizations coated with solder so that the capacitor can be attached by solder reflow techniques. A variety of gold- and silver-bearing conductors are also available,

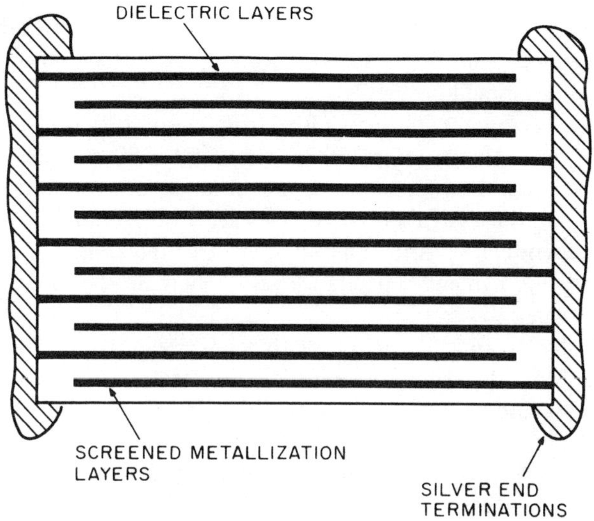

Fig. 9-4. Cross section of a multilayer ceramic capacitor.

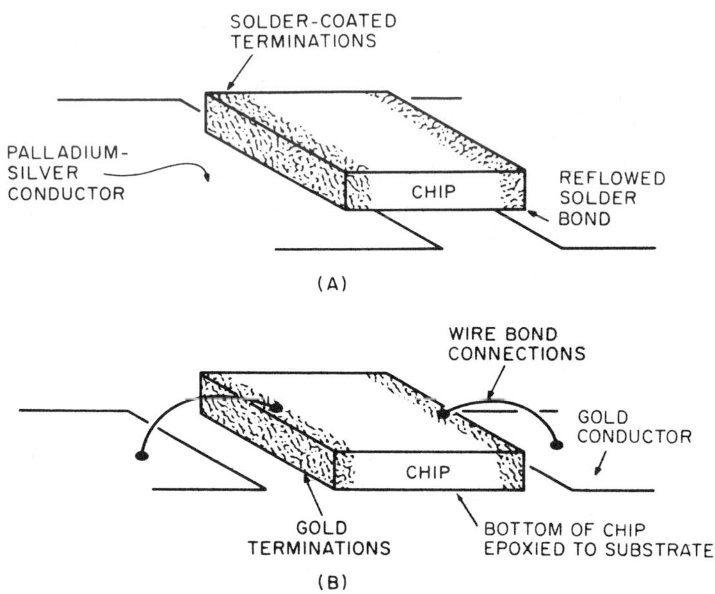

Fig. 9-5. Chip capacitor attachment. (A) Solder reflow. Contact to conductors is made by solder bond. (B) Epoxy bonding. Chip is epoxied to substrate and then connections made by wire bonding.

so the capacitor can be attached by conductive epoxy, resistance welding, ultrasonic welding, and chip and wire techniques. Solder reflow and epoxy bonding are diagrammed in Fig. 9-5.

Values of ceramic capacitors range from about 100 picofarads to about 1 microfarad, in packages ranging from about 50 x 50 mils up to 1/2 x 1/3 inch. Values as low as 10 picofarads are obtainable with suitable selection of dielectric materials. Breakdown voltages are from about 50 volts to about 200 volts, depending on the internal dielectric thickness. Size, of course, increases as breakdown and capacitance values increase. These ceramic capacitors are nonpolar.

Eventually ceramic capacitor sandwiches reach a limiting size, just as the screened films do. There are physical limitations to the thinness of the dielectric layers and to the size to which the package can be increased. The capacitors start to take up too much substrate area to be practical. The solution is to go to a higher volumetric efficiency dielectric. This is provided by tantalum capacitors. Tantalum chip capacitors offer the highest capacitance per unit volume of the currently usable dielectric materials. Tantalum capacitors have been on the electronic market as discrete devices for some years, and now have been followed up by hybrid engineers.

A tantalum capacitor consists of a compressed and sintered slug of tantalum powder with a solid tantalum rod embedded in it. See Fig. 9-6 and Fig. 9-7. The rod is the positive electrode and the slug will become the negative electrode. The slug and rod go through an acid process to produce the dielectric, which consists of a film of tantalum pentoxide (Ta_2O_5). The thickness of the oxide between the slug and the rod will determine the capacitance and the breakdown voltage of the capacitor.

The slug is coated with several layers of manganese dioxide. Then the actual negative connection is made by means of a silver metallized coating

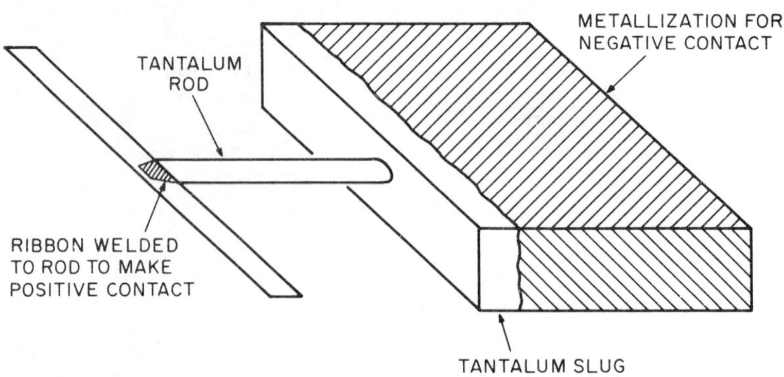

Fig. 9-6. Uncased tantalum chip capacitor.

over the manganese dioxide (MnO_2). The tantalum rod usually has a metallic ribbon of gold welded onto it to form a good electrical connection. In a tantalum capacitor to be used for discrete circuits, the metallization on the body and the imbedded electrode have wires attached to them to form the electrical connections to the outside, and then the whole assembly is molded or potted into a plastic case. For thick film circuit use, the capacitor case is eliminated and the bare slug of tantalum is mounted into the circuit. The silver metallization and the gold ribbon are often solder-coated, so that the capacitor may be attached to the conductors by solder reflow. Alternately, various welding schemes or the use of conductive epoxies may be employed.

Unfortunately, the high volumetric efficiency of the tantalum capacitor is not as ideal as it sounds. Bare tantalum chips can be a lot of trouble if not handled correctly. The problem lies mainly in the electrode that is embedded in the tantalum slug. Movement of the electrode in the slug will lead to changes in capacitance, high leakage, and outright failure. This type of movement can often occur during handling or assembly, so extreme care must

Fig. 9-7. Chip capacitor. Close-up of uncased tantalum chip capacitor. Note tantalum rod protruding from the side, and solder coating on the body for reflow attachment.

be taken in handling these devices. Epoxy strengthening of the embedded electrode will help this problem. Another potential problem is that an uncased tantalum slug is not hermetically sealed. Materials can diffuse into the body of the capacitor. This includes flux from soldering, solvents from cleaning, fumes or other undesirable chemical contaminants given off by some types of packaging materials, such as some epoxies or silicones. This will lead to an increase in leakage across the capacitor and a possible change in capacitance characteristics. These materials can also degrade the oxide dielectric film or can accelerate breakdown through the localized manufacturing of pinholes or impurities. This is not to say that uncased tantalum capacitors cannot be used—they are being used and very successfully—but it is as well to be aware of the possible failure modes.

Actually the solution to the tantalum capacitor problem is much easier than it might seem—simply put a case around it. Several different manufacturers have made available series of subminiature encased tantalum capacitors. Some of them are actually smaller than many of the uncased chips, leading to a volumetric advantage as well. The capacitors are the same type of tantalum chips, except that they are molded into a tiny plastic case and have solder leads or flat ribbon leads projecting from the case. The leads can be attached into the thick film circuit by microsoldering, reflow soldering, ultrasonic welding, or resistance welding. The plastic case does not increase the size very much and it is well worth it in terms of protection for the rod and slug. Subminiature tantalum capacitors can deliver from about 0.01 microfarad to 50 or 100 microfarads in very small sizes, depending on the working voltage needed.

While we are on the subject of chip capacitors, it should be noted that even though some of the capacitors are large, they can be made to take up very little space by running resistors or conductors underneath them. This applies to ceramic sandwiches and discrete tantalums. Since the only conductive parts that connect to the circuit are at each end, there is no reason why

Table 9-1. Comparison of Capacitor Characteristics.

Material	Polarity	Approximate Capacitance	Voltage Rating
Monolithic P-N junction	Reverse biased	1–350 pF	Up to 60 V
Monolithic oxide	Nonpolar	1-500 pF	50 V
Screened	Nonpolar	10 pF to 0.05 μF	50 V
Ceramic	Nonpolar	10 pF to 1 μF	25-100 V
Tantalum	Rod is positive	0.001-200 μF	2-50 V

they should not be used as a bridge to cross conductors and resistors and save some substrate area. A comparison of capacitor chips is made in Table 9-1.

The third type of passive circuit elements used in chip form are chip inductors. Generally inductors and transformers are designed out of the circuit or are wired in external to the thick film circuit, but sometimes there is no way around the inductance problem and coils have to be used. Due to laws of field and magnetic theory, it is very difficult to create large inductance in a small volume. Work has been done with special core materials and

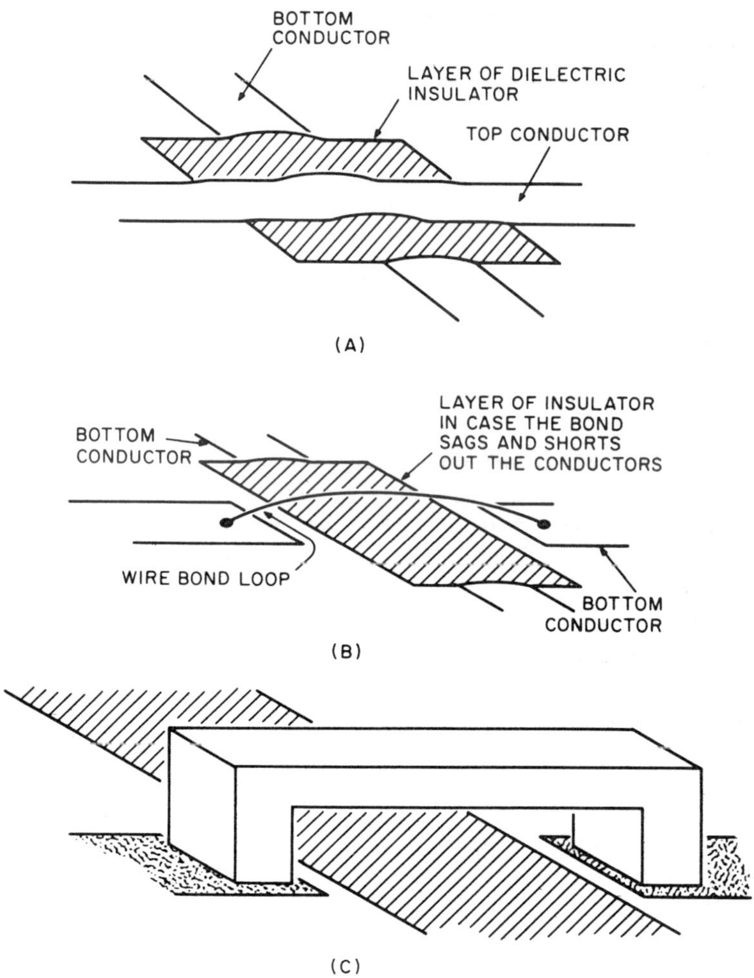

Fig. 9-8. Crossover methods. (A) Screened insulator method. (B) Wire bond method. (C) Discrete short crossover method.

core shapes such as toroids. Discrete inductors have been wound in subminiature form up in the millihenry ranges. These are treated as discrete components and are attached by soldering or welding. Inductors have been fabricated directly onto the substrates by thick film screening and thin film deposition in the form of spiral patterns, but the values are limited. Inductance values of several hundred nanohenries have been fabricated, both in chip form for attachment to a substrate and also fabricated directly onto thick film and thin film substrates. The best way to go as far as inductors are concerned is still to try and design them out of the circuit, either physically by putting them outside the package or by electrically designing around the need for them.

There is one final discrete device that deserves mention—if it can be called a device. This is the discrete short. It consists of a connecting conductive link in chip form. In its simplest form it is a piece of rigid wire bent in the shape of a U. The purpose it serves is to connect two conductor paths across a third one. It is a crossover element.

In practice, connecting two conductors across a third one can be done in three ways. First, a conductor is screened and fired onto the substrate. Then dielectric insulating material is screened on top of it at the point where the crossover is to occur. For best insulation results and to eliminate the possibility of pinholes and shorts, two dielectric screenings would be used. A second conductor is then laid down across the dielectric layer(s) to form the actual crossover, as in Fig. 9-8(A). This requires at least three separate screenings. The second method of forming a crossover is to screen all the conductors at the same time, leaving a gap where one crosses the other. The crossover connection is made by joining the two open ends of the crossover conductor with a wire bond, as in Fig. 9-8(B). This is acceptable, but for utmost reliability, the conductor being crossed should have a dielectric insulating layer screened over it in case the wire link sags and shorts out the two conductors. This method calls for two screenings and a wire link. The third method of forming a crossover is by the use of a discrete short, as in Fig. 9-8(C). This is about the same as the second or wire bond method, except that the discrete short takes the place of the wire bond. The short is self-supporting and will not sag, so no dielectric layer is needed to protect the crossed-over conductor. The discrete short can be reflow-soldered at the same time as other passive components, or it can be ultrasonically or resistance welded.

The variety of discrete devices for thick film circuits is quite large. New materials for capacitors, resistors, and inductors are under constant study, so passive chip devices will see constant improvement in the future.

10 — Active Device Bonding

Semiconductor dice are attached to the circuit following resistor and conductor screening and firing. There are several methods of device attachment: (1) eutectic bonding, (2) epoxy bonding, (3) ceramic carriers or LIDs, (4) micropackages, (5) flip chip technology, and (6) beam-lead devices.

Eutectic Bonding

By far the most common and one of the easiest methods of die bonding is eutectic bonding. Uncased semiconductor chips are used for this. Generally the chips have a thin coat of gold evaporated onto the collector surface to assist in the bonding process, but this is not essential. As mentioned earlier, if the backing is there, it is necessary to ensure that the backing thickness is adequate and that there is no contamination or oxidation underneath the gold layer. This can be worked out with the supplier.

Electrical characteristics of the bare chips themselves are similar to those of the packaged transistor. The characteristics will not be identical because of the effects of package and header capacitances and inductances. In the case of high-frequency chips, the characteristics are often improved by the elimination of these degrading parasitics. The bare chips are bonded directly to the conductor on the substrate, so power-handling capabilities can be quite good.

The actual bonding is done with a special machine called, appropriately enough, a die bonder. The theory is quite simple. The substrate and the chip to be bonded are raised above the gold-silicon eutectic temperature. At this temperature, the chip and the conductor become welded together by the plastic gold-silicon intermetallic. The substrate is cooled and the chip adheres to the conductor. Pure gold or gold-bearing conductors have to be used to form this type of bond. Bonding the chip directly to the conductor is called direct eutectic bonding. It is sometimes called a thermal compression type of bond as it uses the effects of heat and pressure.

Because of the varied nature of hybrid circuit manufacturing, most of the equipment is available on a semicustom basis. The needs of individual processors are so different that most equipment manufacturers only build equipment to a certain point, then finish it off in terms of an individual customer's needs. This customization also applies to die-bonding equipment. Therefore this equipment can only be discussed in quite general terms.

A eutectic die bonder will have the following features: heater stage, die dish, bonding head, optics and lights, manipulators, hand or foot controls, and gas nozzles.

The heater stage is used to bring the substrate up to the eutectic temperature for the actual bonding operation. Usually this is just a heavy block of metal with heater windings or cartridges in it. The thermal mass should be large enough that the temperature stays constant when cold substrates are placed on the block for bonding. The stage temperature is variable by means of a rheostat and/or electric temperature controllers. The temperature is read out by a thermocouple and meter.

There will be some method of holding the substrate on the block during the bonding cycle. The most common method is a vacuum holddown. This is a hole drilled through the block and attached to an external vacuum line. Mechanical clamps may also be used. The heated stage is usually movable in the x,y, and rotary directions to allow accurate positioning of the substrate bonding pad under the collet.

The die dish is a mirror or other flat surface that holds the semiconductor dice before bonding. The dice to be bonded are loaded onto this plate before bonding, so that they will be accessible for the die pickup part of the cycle.

The bonding head contains the collet that is to be used for bonding. This is the tool that picks up the dice, then scrubs them into the conductor. The head moves up and down in a predetermined sequence. This motion is motor driven and controlled by cams and electronic logic for sequencing and accuracy. The head comes down to allow the collet to pick up a die for bonding, raises it out of the way, and then comes back down onto the substrate to make the bond. The head may or may not contain a heater for the collet.

Optics and lights are used to view the whole operation. Manipulators position the head over the die to be picked up and the collet over the substrate at the location where the bond is to be made. Hand or foot controls initiate the pickup and bonding cycles. Usually the actual bonding process parameters are preset on the machine to ensure repeatability and accuracy. Gas nozzles provide a flow of inert gas, usually nitrogen, over the top of the heater block and the substrate during the bonding process. This is to prevent oxidation of the various metallizations.

Some processors like to use a preform for eutectic bonding, though the use of preforms in this case is not essential. This is the choice of the processor. A preform is a small pad of alloy metal placed between the chip and the conductor to make an easier bond. It ensures that there is adequate gold available at the chip-conductor interface. For bonding silicon chips onto gold-bearing conductors, the common preform alloy is gold with 2 to 4% silicon in it.

In practice, the substrate is placed on the heater block at about 400° to 425°C, to ensure that the chip and the substrate are above the eutectic temperature. Vacuum holddown is used to hold the substrate firmly and to pull it into intimate contact with the heater block for best heat transfer. When the substrate is first placed on the heater block, it will cause the block temperature to drop slightly. When a cold chip is placed on the substrate the immediate area will momentarily experience a slight temperature dip. The chip is picked up with a vacuum collet, as shown in Fig. 10-1 (A), and positioned at the desired bonding location. The tool used to pick up the dice is a specially made hollow vacuum chuck, that is machined to very close tolerances. It is usually made of tungsten carbide. Next the collet comes down and

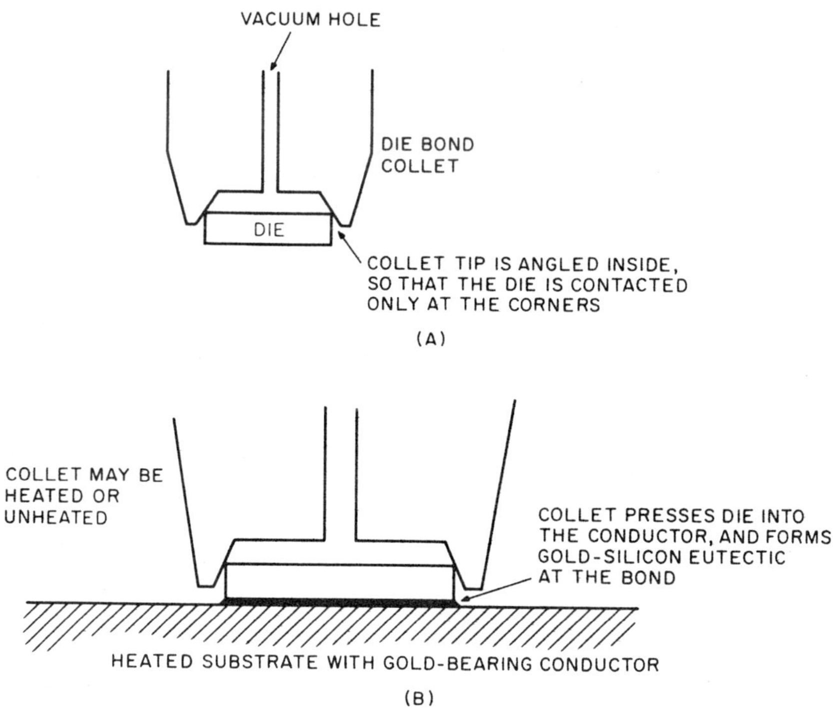

Fig. 10-1. Eutectic die bonding. (A) Die pickup. (B) Bonding.

presses the die into the now-plastic conductor, as in Fig. 10-1(B). Usually a mild scrubbing action by the collet is used to break through the conductor oxide layer and to ensure that the chip is wetted into the conductor. The scrubbing can be a mild back-and-forth or rotary mechanical action or it can be an ultrasonic scrub. Eutectic bonding can be done completely manually without using a die bonder. First, the substrate is placed on a heater or hot plate to bring it to the eutectic temperature and then the die is picked up with a pair of tweezers and rubbed lightly into place. This is not a very fast, accurate, or desirable way to die bond, but it can be done if necessary. At the other extreme are the completely automated systems for die attach. Dice are picked up, positioned, bonded, and indexed to the next substrate by the die bonder.

Heating the substrate for eutectic bonding has one disadvantage. A heating block is used to bring the chip and conductor to the eutectic temperature, so that a fusion can occur. When the bond is being made, a darkening of the conductor around the chip will be seen. This is some of the silicon dissolving into the molten gold to form the bond. This is the way it should be. But now the operator goes on to the next bond. The next chip is picked up, placed in position, and scrubbed into the conductor. Meanwhile, the first chip is still sitting at the eutectic temperature too. The gold is still dissolving silicon from the chip and, since the interface is still plastic, the chip is not anchored firmly to the substrate. If the substrate vibrates, the chip can float around on the conductor and become misaligned.

For better results, eutectic bonding should be done slightly differently. The eutectic temperature is around 370°-385°C. The substrate heater is set a little below this, about 360°C. Then the vacuum collet that picks up the dice is independently heated to provide a temperature a little above the eutectic temperature. The eutectic bond will still form because of the extra heat supplied to the substrate and chip by the collet during bonding. After the bond is made, the substrate and chip will fall back below the eutectic and the chip will be anchored to the conductor. A concentrated infrared or hot gas heating beam, supplied to the bonding area only, is also in use. This also serves to heat only the local area where a chip is being deposited. The actual temperature for the heated collet will have to be determined experimentally. It depends on the substrate temperature and thickness, the die size, the type of scrubbing action, and the conductor type. The way to determine the correct temperature is to raise the collet temperature a few degrees at a time, while making bonds, until good bonds are being made. Test the bonds by trying to shear off the die from the conductor. A good bond is made when the die cracks or chips or the conductor breaks loose from the substrate before the bond breaks. If the conductor does come loose, the substrate cleaning procedure should be re-examined to see if poor adhesion is occurring. The collet should be run at as low a temperature as possible while still making good bonds. This will lessen the collet oxidation and lengthen the life of the collet.

It is preferable to bond directly without the use of a preform. The preforms are quite inexpensive, but their use involves an extra handling operation and a bond with a preform is harder to inspect. The preform involves one extra bond interface and can lead to hard-to-detect failures. Instead of a preform, some of the soft solders are sometimes used and the dice eutectically bonded to presoldered pads. A low tin-content solder or a gold-bearing solder must be used to prevent the gold backing on the chip from leaching into the solder.

The maximum die size that can be handled conveniently by eutectic bonding is around 50 x 50 mils. A larger size leads to some problems with die heating and can cause cracking and chipping during handling and bonding. For larger sizes, many processors turn to epoxy bonding.

There are many opinions on the subject of the correct atmosphere for eutectic bonding. Can eutectic bonding be done in normal room air? The problem here is oxidation. At the die bonding temperatures, the chips and conductors will oxidize quite rapidly and make bonding difficult. This is one reason for scrubbing the die into the conductor—the abrading action helps break through the oxide layer and seat the chip. To combat the oxidation problem, a nitrogen atmosphere is often used. Nitrogen or another inert gas is flowed across the substrate and the heater block to cut down on oxide formation—a good practice, but not essential. The best recommendation is that if oxidation is a problem, a nitrogen cover should be used. To further cut down on oxidation, the dice themselves should be stored under nitrogen before use to prevent oxides from forming on the gold back or the bonding pads.

Epoxy Bonding

Epoxy bonding is commonly used—some large companies use it exclusively for all die sizes. It is one of the best ways to handle the larger die sizes. There are two types of epoxy bonding—conductive and nonconductive. The nonconductive uses an ordinary epoxy to simply glue the chip to the substrate. Then all the electrical connections are made to the top surface. The epoxy bonding sequence is shown in Fig. 10-2.

The conductive epoxy method is the more common. The epoxy used has a gold or silver filler. When cured, the gold or silver forms an electrically conductive path from the conductor to the bottom of the chip. Copper- and aluminum-filled epoxies are available, but they are not used for die bonding. There is some controversy as to whether gold or silver epoxy should be used. Some people use gold only because of the silver migration problems. However, silver does work very well and some very large companies are using it

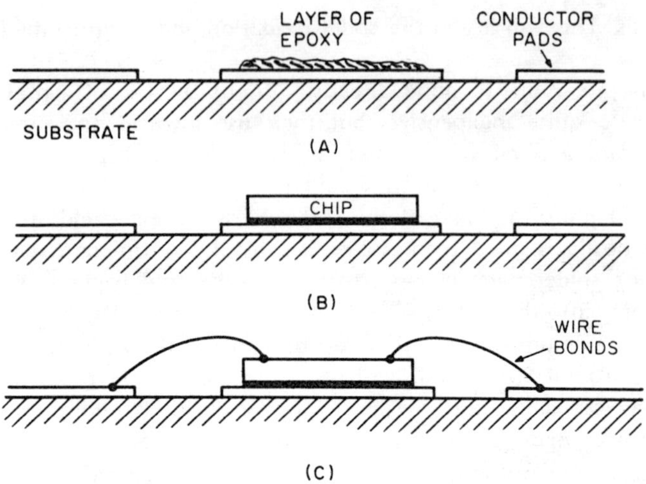

Fig. 10-2. Epoxy bonding sequence for dice. (A) The layer of epoxy is spread onto the conductor at the bonding site. (B) The transistor die is pressed into the epoxy. The epoxy is then cured. (C) Wire bonds are added to the base and emitter contacts.

successfully. As mentioned before, silver migration only takes place in the presence of moisture. A hermetic package will eliminate the moisture and make the migration negligible. Silver epoxy is several times cheaper per ounce than gold epoxy.

Die-bonding epoxies are two-part, low-temperature curing epoxies consisting of a resin and a hardener. Both room-temperature and elevated-temperature curing epoxies are available—the higher temperature ones will cure much faster. Manufacturers' specification sheets should be followed for processing instructions. Single-part epoxies that cure at elevated temperatures are available, but are not as frequently used. The epoxy is applied with a needle, toothpick, hypodermic syringe, or any other convenient dispenser. For production use, air-operated syringes that will meter out an exact amount of epoxy each time, are used. The desired amount of epoxy is deposited on the pad or substrate and the chip positioned with a pair of tweezers or a vacuum pickup. The chip is then pressed into place with a toothpick, plastic tool, or plastic-coated metal tool. The tool used for this should be inert, such as teflon, and have a soft die contact surface to prevent scratching the transistor.

Epoxies doped with semiconductor donors or acceptors are sometimes used to form (or prevent formation of) additional P-N junctions as part of the bond if necessary.

Ceramic Carriers

Bare dice can be an advantage or a disadvantage, depending on the process and desired results. There are two serious disadvantages. First, bare dice are difficult to handle. They are very small and can be easily damaged and scratched. A carelessly used pair of tweezers can ruin dice very quickly. Second, they are difficult to test. Tooling for testing and handling a 15 x 15 mil semiconductor die in quantity is not a very easy task.

Electrical testing of dice is also a problem. Testing on uncased chips is limited to several simple dc tests, usually breakdown voltage, dc beta, and perhaps leakage testing. The reasons for this are twofold. The first is purely mechanical. To test a chip, mechanical probes have to be pressed into the bonding pads to make electrical contacts. If this is done more than a couple of times, physical damage to the chips will occur. Second, the tests are not very reliable. Ac testing cannot be done because the package will make a difference to the readings. Header and lead capacitances and the package material will affect the ac values. Since uncased chips are light-sensitive for the most part, precautions must be taken to account for this. Leakage values, especially, will be radically affected by light shining on the device during probing. The values obtained by probing are not the final values that will be obtained in the circuit. The high temperatures used for eutectic bonding will degrade the chip parameters to some extent. Values of dc beta may drop by a third or a half in extreme cases. Breakdown voltages can decrease and leakage currents increase at the same time. Transistor and diode chips were designed to be bonded singly into single cans. This takes only a few seconds. Bonding of several active devices onto a thick film substrate can take several minutes or more. These relatively long times at bonding temperatures of 360° to 400°C can seriously degrade the transistor performance. This means that no conclusive tests can be made until the device has been bonded onto the substrate—and then it is often too late.

Some processors prefer to buy the semiconductors in wafer form and test and sort them themselves, before scribing and breaking the wafer apart. This can create handling problems. If the dice are being sorted for some particular parameter, additional handling problems are created. High-speed automated equipment is available for device sorting, and is in use at the large semiconductor houses. For most thick film processors, the initial cost is too high and the quantity of chips tested is not sufficient to justify the cost. Another chip problem can occur following final assembly. The wire bonds to the chip are fragile and easily broken during handling if care is not taken.

To help solve these problems, micropackages called ceramic channels or leadless inverted devices (LID)* are used. There are various different versions

* LID is a trademark of Amperex Electronic Corporation.

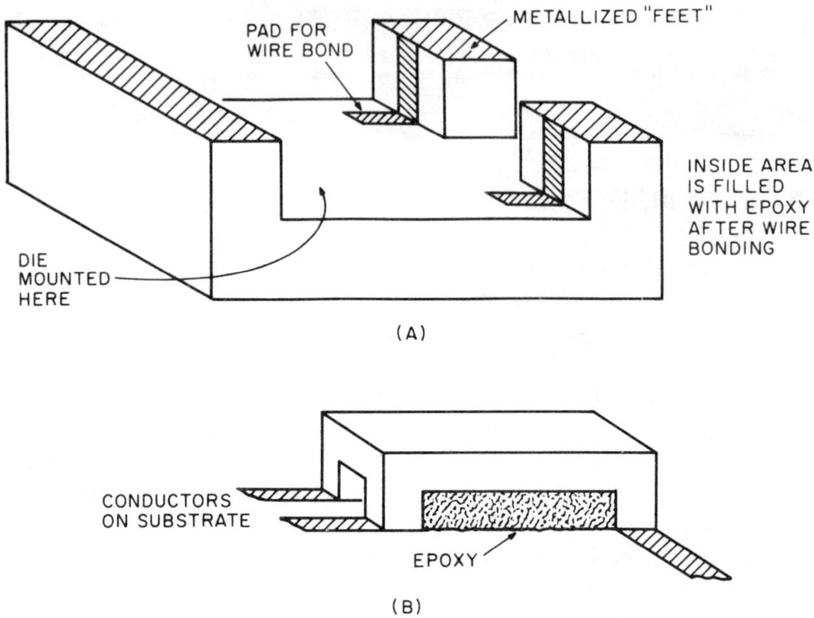

Fig. 10-3. Ceramic channel or leadless inverted device (LID). (A) Transistor or component is mounted into the LID. (B) Device is inverted, and metallized pads are adhered to the conductor by epoxy, solder reflow, or ultrasonics.

of this and the carriers come in various geometries and sizes to accomodate many different chips, but all are basically alike. They consist of a ceramic base for mounting the component, with raised portions on which there is metallization to make the external connections. A typical LID and bonding process is shown in Fig. 10-3. The dice are bonded to the ceramic base of the LID and wire bonds made to conductive paths that go to the external metallization. Then the chip and the wire bonds are all protected with epoxy. At this point the devices can be treated as discrete semiconductors. The transistor can be ac- and dc-tested, since the parameters will be the same when the devices are attached to the substrate.

For attachment to the circuit, the devices are inverted and the metallization on the legs becomes the circuit connectors. These packages can be attached by reflow soldering, conductive epoxy, or ultrasonic welding. The carriers have the advantages of protecting the chips and the bonds, and they can also be used to bridge a path on the substrate, leading to good packing densities. The use of ceramic channels is not restricted to transistors. Diodes, resistors, and capacitors can also be handled by this method.

Circuit elements can be purchased already packaged or the ceramic carriers can be purchased and the desired elements installed in-plant. The disadvantages include large size, the additional cost of the carrier, and the equipment for attaching the carriers to the substrate. If devices already installed in carriers are bought from outside vendors, the cost will be higher than bare chips because of their processing time. If handled in-house, there is the internal cost of time and equipment for installing the chips and for testing and preparing the final package.

Micropackages

Several manufacturers offer their standard devices in smaller than normal packages. The packaging may be a small ceramic chip with leads attached and a drop of epoxy on it or it may be a tiny molded plastic package with leads sticking out. Whatever the method, the idea is to protect the bare chip from the outside world and bring the connections out as relatively sturdy leads. These packages can be quite small. Transistors are available in discrete packages that are not much bigger than the head of a pin. The leads are more fragile than on a regular transistor package and should be handled with some care, but no unusual precautions are necessary. The leads are attached to the substrate by conventional means: reflow soldering, microsoldering, welding, or ultrasonic bonding.

Package size is somewhat larger than will be obtained by chip-and-wire techniques. However, as in LID devices, the active semiconductor portion is protected from the possibility of mishandling and other abuses. These subminiature devices can be fully tested before use.

Flip-Chips

One of the most common failure modes in thick film circuits is the failure of the wire bonds. Weak bonds, stressed wires, overbonded wires, and other defects lead to most of the failures in chip-and-wire circuits. The logical solution is to eliminate the wire bonds altogether. This is the approach taken by flip-chip and beam-lead devices, though they go about it in slightly different ways.

Flip-chips consist of bare semiconductor elements. On each of the bonding pads there is a raised metallic bump of solder, aluminum, or copper. A flip-chip die and a bonded die are shown in Fig. 10-4. The bumps correspond to conductive land areas on the substrate. For bonding, the chip is flipped over (hence the name) so it is actually face down, the pads and bumps are lined up through special optics, and then the chip is attached to the substrate by metallurgically joining the bumps to the land pads on the substrate. This is done by heating and melting the bumps in the case of solder or by ultrasonic scrubbing in the case of aluminum.

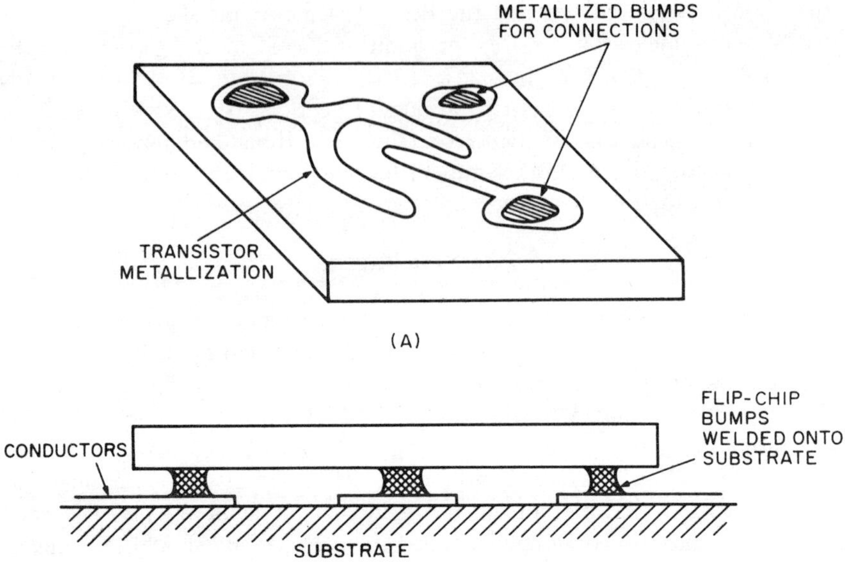

(B)

Fig. 10-4. Flip-chip attachment. (A) View of the bottom of flip-chip transistor. (B) Transistor inverted and bonded.

Flip-chips have the advantage of making all the connections at once and of eliminating two wire bonds per connection. Since the connections are underneath the chip, some layout space is saved. No wire bonding pads are needed.

As usual, along with the advantages there are also disadvantages. The first is cost. Flip-chip equipment is necessarily more expensive than conventional die attach equipment. The flip-chip bonder must be very precise and able to line up all the bumps and the pads accurately for the chip to be put down and bonded in exactly the right place. And there lies the second problem with flip-chips. Since the connections are all under the chip, there is no easy way of seeing a weak bond after it is made. Most of the problem here is the fault of the substrate. Chapter 4, on substrates, deals with the problem of local camber. The local variations in surface flatness can be quite large in some areas of the substrate. This plays havoc with flip-chips. If the chip is bonded in a true horizontal plane and the substrate is warped, either locally or overall, then some chip bumps will be flatter than they should be and some will not be collapsed as much as they should be (Fig. 10-5). This, of course, leads to weak bonds and possible failures and there is no way of seeing through the chip to tell that this has happened. There are two solutions to this particular problem. One is to buy better substrates, such as glazed substrates or ones with a higher alumina content. The other solution is to use a profilometer and check the flatness of the substrate in the area where the chip will be bonded. The first solution is expensive in terms of a higher cost of materials. The second solution is expensive in terms of time, labor, and the cost of rejected material.

Another major disadvantage of flip-chips is that they are limited in power-dissipation capabilities. The chip is slightly raised from the substrate by the bonds and is not in intimate contact with the substrate as is a eutectically mounted die. Thus the power dissipation cannot be as great.

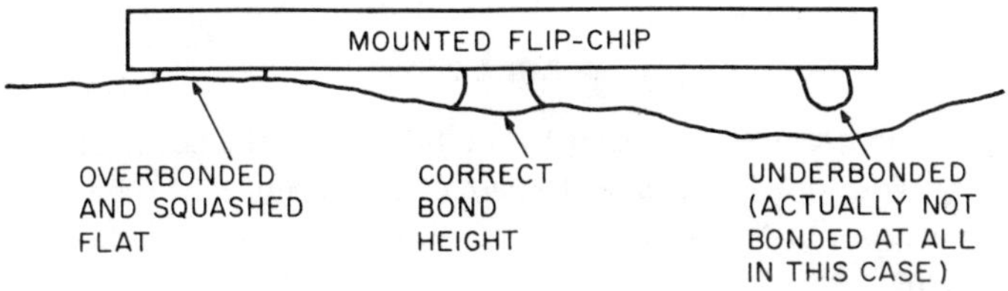

Fig. 10-5. Exaggerated view of local camber problem encountered with flip-chip bonding.

In spite of their disadvantages, flip-chips may be a dominant technology in the future. Much work is being done to improve the metallurgy of the connections and the equipment used for bonding. In spite of present problems, flip-chips look like they have much to offer in the way of reliability and ease of bonding.

Flip-chip bonding equipment is almost identical to standard eutectic die attach equipment. There is a die dish to hold the dice to be mounted, a collet to pick up the dice for transfer to the substrate, and a movable stage with clamps or vacuum to hold the substrate during die attach. The main feature that makes a flip-chip bonder different from a conventional machine is the optics. In flip-chip attachment, the operator has to see the underside of the flip-chip to see the metallization bumps, and at the same time has to be able to see the metallization pattern on the substrate to know where to locate the bumps on the substrate metallization. This calls for the use of mirrors, half-silvered mirrors, and prisms to superimpose the two views. The mechanical manipulation of the bonding head and the substrate also has to be more sophisticated to allow accurate alignment of the flip-chip on the substrate pads.

Flip-chip bonders come in three different types: ultrasonic, thermal compression, and solder reflow. The ultrasonic type uses an ultrasonic generator the same as an ultrasonic wire bonder (discussed in Chapter 11). The ultrasonic generator feeds a high-frequency electrical signal into an ultrasonic horn or transducer that converts the electrical energy into a mechanical action through the collet and scrubs the flip-chip bumps into the bonding pad. This scrubbing action creates enough heat to form a metallurgical weld of the bumps onto the pads on the substrate.

The thermal compression type of bonder uses the effects of heat and pressure to form the metallurgical bond between the chip and the substrate. Either the substrate-holding stage, the collet holding the flip-chip, or both are heated. The solder reflow bonder is used with chips that have solder bumps. The heat melts the bumps and reflow solders the device onto the substrate metallization.

Beam Leads

The other wireless approach to die bonding is the beam-lead approach. Beam-lead devices are very similar to bare chips, except that they usually have better passivation. Instead of having metallization patterns on the top, as does a conventional die, the beam-lead devices have their outside connections coming out as thin, flat leads or "beams," projecting out from the bottom of the die. Beams are typically only 2 to 4 mils wide. The beams are formed as part of the device during wafer fabrication and the wafer is then etched apart to give the separate devices. A beam-lead transistor is sketched in Fig. 10-6.

Beam-lead transistors and monolithic integrated circuits are bonded to the substrate metallization by thermal compression techniques, resistance, and parallel gap welding and ultrasonic heads. The device is picked up with a vacuum collet and positioned over the desired bonding area. The collet descends and applies heat, pressure, or ultrasonic vibration to the beams to attach them to the conductors. The beams themselves are quite fragile and different bonding techniques are used to try and maximize bond strength while minimizing damage to the devices. Beam adhesion to the conductors is usually pretty good. Failure occurs most often where the beams are connected to the silicon device.

Similar to flip-chip bonding, uneven substrates will produce uneven beam bond pressure and variations in bond strength, leading to possible bond failures. Also similar to flip-chips, beam-lead technology could become a major bonding method. The devices are all fully passivated and protected. The bonds are all made at the same time that the device is physically mounted on the substrate, leading to a decrease in processing time by eliminating the wire bonds. The bonds are much more rugged and reliable than chip and wire bonds.

Beam-lead bonders are very similar to standard eutectic bonders. There is a die pickup dish, a substrate-holding platform, mechanical or electrically operated manipulators to position the substrate, and a collet that picks up the device and then bonds it onto the substrate metallization.

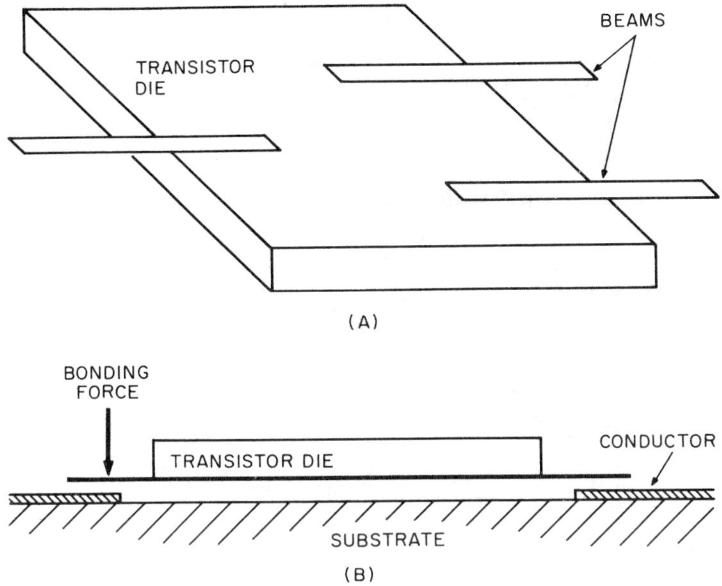

Fig. 10-6. Beam-lead attachment. (A) View of the bottom of the beam-lead transistor. (B) Inverted die in position for bonding.

The bonding is performed by thermal compression techniques or by ultrasonic energy scrubbing the beams into the conductors. The ultrasonic scrub bonding does not seem to have gained much popularity, because of the fear that the high frequency energy may damage the chip. The thermal compression technique is generally used. The bond is made by the application of heat and pressure to the beam and the conductor. To ensure that all the beams receive adequate bonding force, several different methods are used to apply the pressure to the beams. This is not so critical on a diode or transistor that has only two or three beams, but on a twenty- or forty-lead integrated circuit it is very important to make sure that each beam makes a good bond. One method rocks the tool back and forth and side to side on the beams. Another method rocks the tool in a spiral motion, so that each bond is made in sequence. A method that is becoming quite popular is called compliant bonding. This uses a deformable material, such as aluminum foil, between the tool and the beams. The foil is deformed by the tool pressure and helps spread the bonding force more evenly over the beams. The deformation of the foil absorbs any slight irregularities in substrate camber, tool finish, and beam dimensions, yet it allows the tool pressure to be transmitted to the beams to make the bond.

Like flip-chip bonders, the beam-lead bonders have quite sophisticated optics. On a multilead integrated circuit with beams projecting from all four sides, the alignment problem becomes critical. The beam width might typically be 3 mils on 10-mil spacings. Therefore mirrors, prisms, and half-silvered mirrors have to be used to provide a superimposed image of beams and bonding pads and ensure that the device is correctly located. Manipulating and die-orienting features also have to be accurate and precise.

Comparing the Bonding Methods

The next few paragraphs offer a comparison of bonding methods. The material is somewhat dated for the same reason that any state-of-the-art report is dated. Today's methods and predictions are only valid in light of today's technology—tomorrow's technology will change the picture.

However, in terms of today's technology, eutectic bonding and chip and wire techniques are the most commonly used and are here to stay for a long time. Most manufacturers are willing to make their devices available in chip or wafer form either tested (to a certain degree!) or untested. Many of the larger distributors are handling chips as off-the-shelf items. More devices are available in chip form than any other form, and chip and wire techniques offer flexibility, small substrate space requirements, ready delivery, and low cost.

Eutectic bonding is more common than epoxy bonding. Generally it is better to eutectically bond as this provides the best metallurgical bond and very low electrical collector resistance. Epoxy is being widely used, but can have the following disadvantages:

1. Heat limits the operation of the circuit to the degradation point of the epoxy.
2. Epoxy often gives a higher collector saturation resistance than a eutectic bond.
3. Chips have been known to "slide" on the epoxy and break the bonds.
4. Limited pot life of the raw materials and more difficult to process.

Whether or not these factors are a disadvantage depends on the individual process. There may be heat-sensitive elements present in the circuit that require the use of low-temperature curing epoxy rather than the high processing temperatures of eutectic bonding. Other factors may make epoxy possible where eutectic bonding is not.

Ceramic channel and LID devices are not as readily available as raw chips. Empty LID packages are readily available and users can put whatever they want in them if they feel that the additional processing steps are worth it. These ceramic carriers do take up additional space on the substrate, but have the advantage of being able to bridge conductors and resistors. They also offer the advantages of a fully tested device. Lack of wire bonds leads to increased reliability and handling ease.

Discrete micropackaged transistors are available only in whatever is in a particular manufacturer's line unless a special product is manufactured on arrangement with the manufacturer. These packages can be quite large, thus defeating the purpose of hybrid miniaturization. The leads can be quite large and may take specialized equipment for attachment to the substrate. Whether or not these devices are suitable depends again on the user's needs.

Flip-chips are not, at this point in time, readily available. Most of the major manufacturers are looking hard at this and some have devices on the market. There are many problems to be solved and the availability of flip-chips will not become widespread until some of these are under control. Bonders for flip-chips are available from various manufacturers, though these are necessarily more complex and expensive than ordinary eutectic bonders. Flip-chips do seem to be a reliable chip-joining technique and will probably become more widespread in the future.

The same comments are true of beam-lead devices. Availability of beam-lead transistors and integrated circuits is not very good at this time, although bonders and devices are on the market and are becoming more common all the time. Beam-lead bonding also appears to be a very reliable method of die attachment and is the subject of much study.

11 — Wire Bonding

The resistors and conductors are deposited and adjusted. The transistor dice are mounted. Chip components have been added. There is one more element to be added before the circuit becomes operational—the wire interconnects.

Transistor dice have two wire bonds to be made on the top surface to form the electrical connections to the emitter and the base metallizations. The collector connection is made through the metallization on the bottom of the die, but connections have to be made from the other two pads down to the rest of the circuit. This is done by bonding fine aluminum or gold wires from the base and emitter metallizations to conductor pads on the substrate. Flip-chip and beam-lead devices do not have these connections—their connections are made directly to the circuit during die bonding. A circuit with flip-chip and beam-lead devices may still require some wire bonding. Chip resistors, capacitors, or inductors may have wire bond terminations, and wire bonds may be needed to connect the substrate to the package terminations of the external leads. Wire bonding is done by two techniques: thermal compression bonding and ultrasonic bonding.

Thermal Compression Bonding

As the name implies, thermal compression bonding, or thermocompression bonding as it is also called, uses the effects of heat and pressure to join the bonding wire to the pad. The third variable in this process is the length of time that the heat and pressure are applied. All three are interdependent and must be very closely controlled.

Furthermore, there are two types of thermal compression bond, the wedge bond, and the ball bond. Either a wedge-wedge or a ball-wedge combination can be used to provide circuit interconnects. Both types of bond use the same principle to make a bond. The substrate or header is placed on a heated stage and brought to around 300°C. This is slightly below the temperature at which any of the bonding alloys will melt. For a wedge bond, the wire to be bonded is placed over the desired pad, then a heated tool in the shape of a wedge comes down and presses the wire into the pad. The heated tool raises the temperature of the actual bond enough to allow a bond to be formed. At the interface between the wire and the pad, the two melt together and form a metallic bond. The pressure, the temperature, and the length of time that the pressure is applied are the parameters that control the bonding. Two bonds are formed during each bonding cycle. One is formed at each end of the wire link. The bonds are referred to as the first and second bonds, referring to the order in which each bond is made. A wedge bond can be used for either the first or second bond of a bonding cycle.

Ball bonding is quite similar. The wire is fed through a cone-shaped glass or tungsten capillary (Fig. 11-1) that is heated like the bonding wedge. A small hydrogen flame is passed over the end of the wire. The flame has the dual function of cutting the wire after a bond has been made and of melting back part of the wire to form the ball. When gold wire is melted, the end left hanging from the capillary will draw back up into a ball slightly larger than the capillary opening. For bonding, the capillary with the ball on the end

Fig. 11-1. Typical thermal compression wire bonding capillary. Only the very tip is in focus, due to the extreme magnification used. The hole in the tip is less than two-thousandths of an inch in diameter.

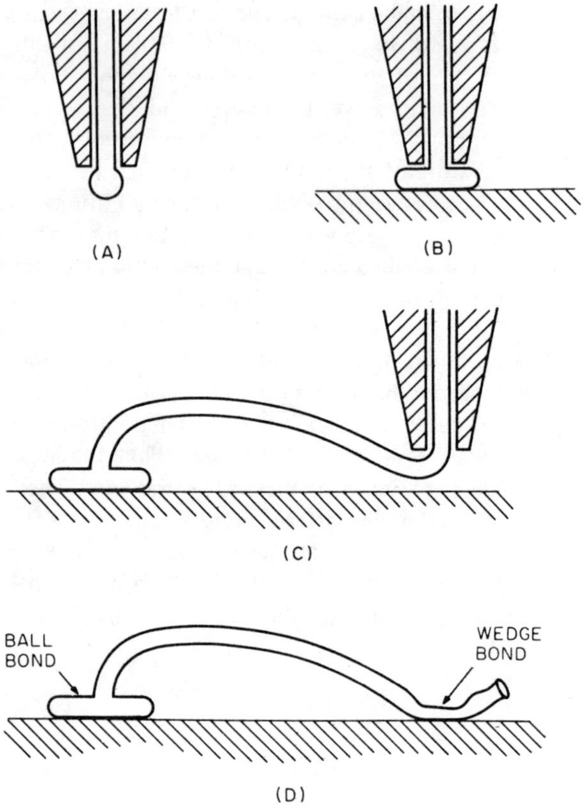

Fig. 11-2. Typical thermal compression bonding. (A) The capillary with the gold ball. (B) The capillary descends to the substrate and bonds the ball to the substrate with heat and pressure. (C) The capillary rises and feeds out wire, then descends to make a wedge bond. (D) The wedge bond is made, the capillary rises, and the ball is reformed for the next bond.

descends onto a pad and metallurgically bonds the ball into the pad by applying heat and pressure. A ball bond can only be used to make the first bond of a bonding cycle; the second bond has to be a wedge bond. The capillary is also used to make the wedge bonds. The side of the capillary tip takes the place of the bonding wedge. Typical ball bonding is shown in Fig. 11-2. A photograph of ball bonds is shown in Fig. 11-3.

The use of a bonding wedge restricts the bonding tool motion. Because of the direction of wire feed for a wedge, wedge bonding is limited to a front-to-back motion. After the first bond is made, the substrate has to be moved straight forward under the tip so that the wire will feed out correctly.

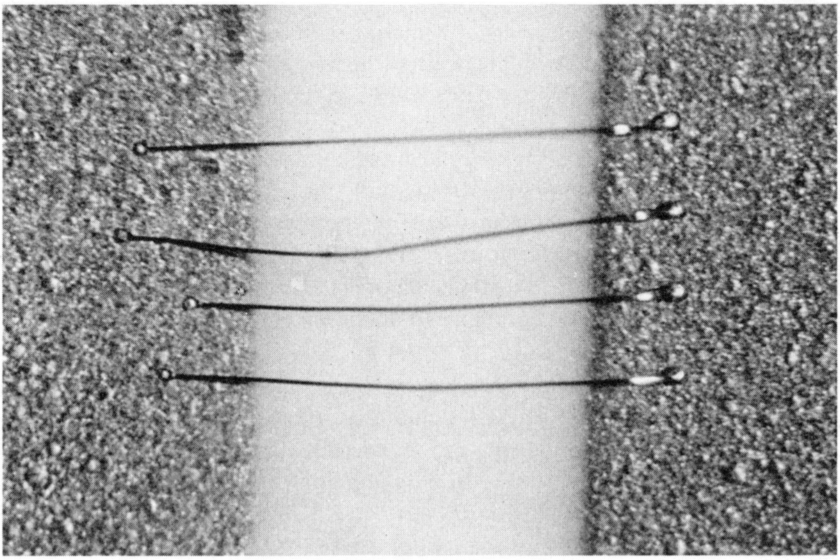

Fig. 11-3. Typical series of thermal compression ball bonds. Note the ball bond made first, on the right, and the tailless wedge bond, on the left.

The second bond is made in a straight line behind the first. The use of a capillary makes this unnecessary. Since the wire feeds through the middle of the capillary, any part of the capillary tip can be used for the second (wedge) bond. The substrate can be manipulated in any direction from the first bond and the second bond can still be made, a far more flexible arrangement.

A thermal compression wire-bonding machine is similar to a eutectic die bonder. Both use the effects of heat, pressure, and time to make a metallurgical weld.

The bonding tool of a thermal compression bonder is called the capillary. It looks like an inverted cone with a tiny hole at the very tip. Figure 11-1 shows the end of a typical bonding capillary. The wire is threaded through the hole in the tip. It is this tip that is used to force the wire and the conductor or die metallization together under conditions of heat and pressure to form the bond.

The substrate is held onto a mechanical stage either by clamps or by vacuum, and manipulated by mechanical linkages to align the capillary over the bonding pad. The capillary is moved up and down to the desired bonding height and the bonding cycle is initiated. The motion of the capillary coming down and making a bond, then rising up and returning to rest, is controlled by cams, electronic logic, and preset controls. The ball is formed by a hydrogen flame that heats the end of the bonding wire momentarily and melts it. Capillary action pulls the melted part up to form the ball. The hydrogen flame is usually either ignited momentarily and shoots out at the wire, or a permanently lighted flame moves in an arc across the wire to melt it.

The heat required to make the thermal compression bond is supplied by heating the substrate-holding block, as with the eutectic die bonder, or by using a heated capillary or by heating both. Temperature levels are adjustable by rheostats and electronic controllers. The substrate-holding stage is movable in the x, y, and rotary directions by mechanical manipulators to allow accurate positioning of the bonding pad under the capillary.

The bonding force is usually preset by a series of dead weights on the bonding tool arm or by some mechanical pressure system. A microscope on the bonder is used to view the bonding operation and can be used for checking and inspecting the process. A cover gas may be used to provide a nonoxidizing atmosphere for the bonding operation. Nitrogen, or other inert gas, is flowed out through nozzles onto the substrate-holding block. The reason for this is that the heat supplied to the substrate for bonding may form oxidation on the metallizations, such as the aluminum on the dice, and prevent proper wire bonding.

A typical thermal compression bonding cycle might run as follows:

1. The substrate is placed on a heated chuck and held in place by vacuum.
2. The machine is recycled to form a ball on the end of the capillary.
3. The substrate chuck is manipulated to place bonding pad underneath the bonding tool.
4. The capillary descends to the first search position. In the search position, the bonding tip comes down very near the substrate to allow final positioning in the x and y directions and height adjustments to locate the bonding tool over the pad to be bonded.
5. The capillary descends to the pad, applies heat and pressure for a predetermined length of time to form a ball bond.
6. The bonding head returns to the rest position, feeding out wire through the capillary at the same time.
7. The substrate chuck is positioned for the next bond. Wire feeds out through the capillary.
8. The capillary descends to the second search position for final adjustment.
9. The capillary descends to the pad and makes a wedge bond, applying heat and pressure again.
10. The bonding head rises up to the rest position again. As it does, the hydrogen flame swings across, cutting the wire to complete the bond and at the same time melting the wire sticking out through the capillary back into a ball to prepare for the next bond.

There are many variations that are used and the above sequence is only representative. Wedge bonds can be used for both bonds using a bonding wedge. The wire may be clamped and broken after the second bond, before the wire is reformed into a ball by the flame. The capillary-substrate heating combination has many variations. And of course all the functions are mechanized so that the pressure, heat, and time are precisely controlled and repeated, and the wire feed is automatic, and so forth.

Ultrasonic Bonding

Similar to thermal compression bonding, ultrasonic bonding also uses a melting effect at the wire-pad interface to make a bond. In the ultrasonic method, the heat used to melt the two members is created by the friction of the wire being rubbed very rapidly against the pad. An electronic oscillator is used to create ultrasonic energy, usually at 60 kilohertz. This frequency is amplified and then fed into a special transducer called the horn that converts the electrical energy into a mechanical back-and-forth scrubbing action. Ultrasonic wire bonding can either be done with a wedge or a capillary. The bonding wedge is used almost exclusively, though ultrasonic gold-ball bonding with a capillary is being done. As with thermal compression bonding, the wedge limits the bonding cycle to a back-to-front motion. The capillary is omnidirectional.

The ultrasonic bonding wedge is located right at the tip of the horn. The wedge does the actual scrubbing of the wire into the bonding pad. For wire-feed convenience, the wire is fed through a hole in the tool and then out under the tip (Fig. 11-4). When the second bond is made, a clamp behind the tool closes and pulls the wire to break it at the tool. When the bonding wedge rises up back to the rest position, the clamp pushes the wire back under the tip again to be ready for the next bond. The clamp stays open between the first and second bonds of the cycle to allow the wire to feed out through the tool. The clamp usually has an override that allows it to be kept open to provide a series of connected bonds called stitch bonds. A typical bonding head is shown in Fig. 11-5.

Like die-bonding equipment, wire bonders are different in their operation and construction, but there are general features that can be discussed. Ultrasonic wire bonders consist of two parts. The first is the ultrasonic energy generator and the transducer to convert this to a mechanical scrubbing action. The second half consists of the rest of the machine, which is a mechanical manipulation system to bring the wire and the substrate into contact with the ultrasonic energy.

Three factors determine the ultrasonic bond. They are the force of the bonding tool on the bond, the amount of ultrasonic energy fed to the tool, and the length of time that the energy is fed to the tool.

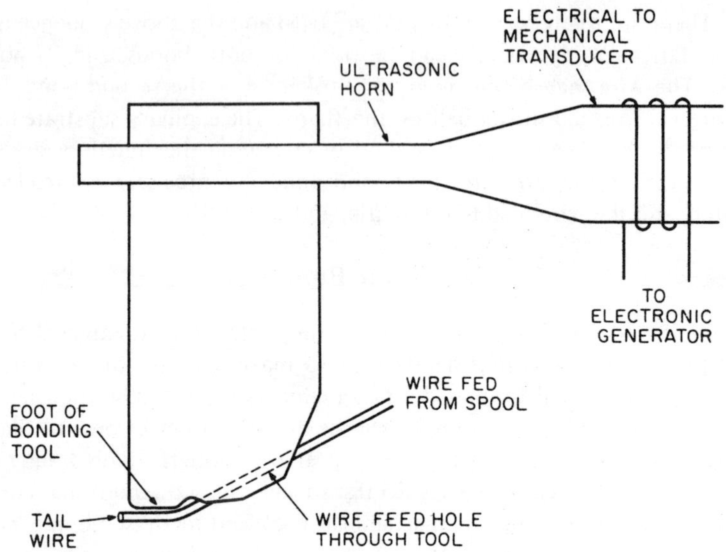

Fig. 11-4. Ultrasonic bonding wedge (*not to scale*).

The force of the tool is controlled by a series of calibrated dead weights on the bonding arm or by counterweights and lever action. Both methods provide a constant force of the tool on the wire and substrate.

The energy and the time are controlled by the electronics of the system and are usually variable. The energy is produced by an electronic oscillator, and is then amplified, put through electronic gating to provide accurate and repeatable time durations, and fed into the ultrasonic horn or transducer. This is a tuned electromechanical resonator that converts the amplified electronic signal into a mechanical vibration that scrubs the wire into the metallization to be bonded. The energy level and time duration are variable to compensate for different wire sizes and compositions, different conductors, and other variable factors. The settings will have to be determined experimentally by the user for different materials.

The rest of the machine consists of mechanical manipulators to put the wire into contact with the metallization so that bonding can take place. There will be a moving stage with either vacuum holddown or mechanical clamps to hold the substrate steady for bonding. This mechanical stage is movable in the x, y, and rotary axes to allow placement of the bonding area underneath the bonding tool. The bonding tool will have to move up and down to allow positioning of the substrate and then actual bonding of the wire. It also has to move up and down to compensate for the various different heights of dice, packages, and substrates. This motion may be controlled manually or by motorized action. There will, of course, be a microscope and stage lighting to

Fig. 11-5. Ultrasonic bonding head. The bonding wedge is located over the bonding site on the substrate. Note the wire clamp behind the bonding tool with wire fed between the jaws.

allow observation of the bonding cycle, and to line up the tool and the bonding areas. The bonding wire is fed from a spool down through or underneath the tip of the bonding tool. Mechanical clamps of some kind are used to clamp the wire and pull off the wire tail after the second bond has been made.

The actual bonding cycle is semiautomatic. Bonding parameters are preset and then controlled by cams and electronic logic. After the operator has lined up the bonding tool over the desired bonding location and initiated the bonding cycle, the machine will complete the cycle by itself.

The ultrasonic cycle is very similar to the thermal compression cycle:

1. The ultrasonic wedge descends to the first search position. The substrate receives final positioning and height adjustments.
2. The wedge descends and scrubs the wire into the bonding pad to make the first bond. The wire clamp opens and wire feeds out from the spool as the wedge rises up to the rest position.
3. The substrate chuck is manipulated to bring the second bonding site under the bonding tool.
4. The bonding wedge descends and scrubs the wire into the pad. Before the tool rises up off the substrate, the wire clamp closes and pulls the wire backwards to break it at the back of the tool, just behind the bond.

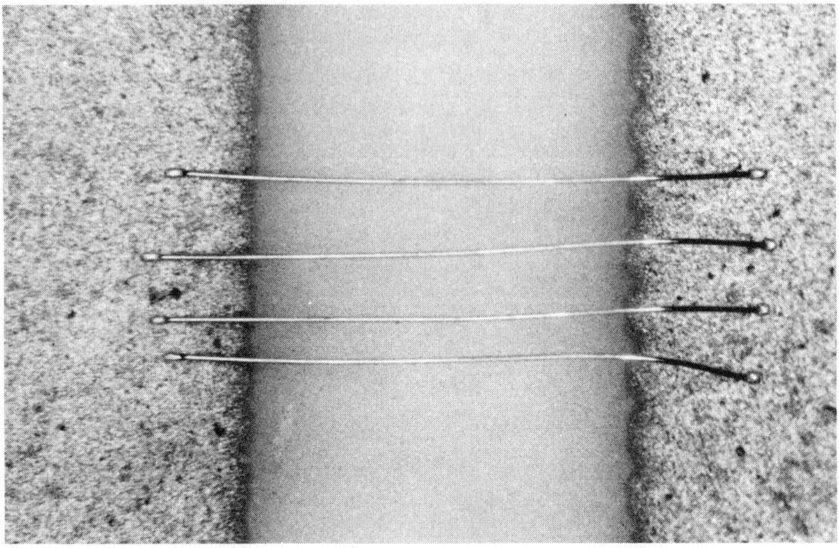

Fig. 11-6. Close-up of ultrasonic wire bonds going from one conductor to another.

5. The bonding wedge rises back up to the rest position. As it does so, the wire clamp comes forward again pushing the end of the wire back under the bonding tip, ready for the next bonding cycle. Typical ultrasonic wire bonds are shown in Fig. 11-6.

The Wire

The wire used for wire bonding consists of two basic types—gold wire and aluminum wire. Ball bonding, either thermal compression or ultrasonic, is done with gold wire. The gold has to be high purity (99.99%) and is usually annealed for better handling characteristics. Gold is used because it melts and forms a ball easily when cut by the hydrogen flame. Aluminum wire will not do this. Aluminum wire bonded with a capillary tip has to be broken after each complete bond and the tail bent back under the capillary tip. There are applications where this may be useful. Thermal compression wedge bonding is usually done with gold wire, although aluminum wire can be used.

Conversely, ultrasonic wedge bonding is usually done with aluminum wire, although gold wire can be used. Pure aluminum wire is soft and not very strong, so it is rarely used. Instead, an alloy of 99% aluminum and 1% silicon is used to increase strength and hardness. One percent magnesium can also be used instead of 1% silicon. One reason why the aluminum ultrasonic bonding system has received such wide acceptance is that it eliminates the gold-aluminum plague problems which have occurred when using gold thermal compression bonding. Another reason is that aluminum is superior for ultrasonic bonding. Part of the key to ultrasonic bonding is that the motion of the ultrasonic wedge scrubs the wire into the bonding pad, abrading away any contamination or oxidation as it goes. The oxide that readily forms on the aluminum wire makes a very good abrasive compound to help in this scrubbing and cleaning action. Gold is fairly soft and does not have these oxides. Ultrasonic tool life of up to 500,000 bonds can be expected, depending on operating conditions.

Wire sizes for both thermal compression and ultrasonic bonding are the same. The most common size for general usage is 1 mil. Of course, 0.7 mil wires and smaller are being used in the lab for research work, but 0.7 mil is about the minimum that should be used for production work. A better size is 1 mil. Larger wires of 3 and 5 mils are being used for special purposes. Greater current-carrying capacity, such as for power transistors, makes this necessary. Wires up to 15 mils are used for some high-power applications.

The temper (breaking load and elongation) of the wire used is quite important. Typical specs for 1-mil-diameter wire are shown in Table 11-1.

Table 11-1. Wire Temper.

	1% Silicon-Aluminum		Gold	
	Breaking Load (grams)	Elongation (%)	Breaking Load (grams)	Elongation (%)
Hard	15	½-1	15	1
Stress-relieved	10	1-3	8	2
Annealed	6	3-7	7	7
Dead soft	6	7-10	6	10

These figures are only representative—the actual wire characteristics will depend on the specific wire manufacturer. The type of wire used will depend on the individual processor's needs. Some people find one temper works much better for them than another temper. The two most commonly used are the hard and the stress-relieved.

To sum up, both thermal compression and ultrasonic wire bonding are in heavy use. Thermal compression bonding most commonly uses gold wire and a capillary tip to do ball and wedge bonding. Ultrasonic wire bonding most commonly uses 1% silicon-aluminum wire and a bonding wedge.

A Comparison

Some advantages and disadvantages of thermal compression bonding compared to ultrasonic bonding are:

Ball bonding generally makes a smaller bond than ultrasonic bonding. Bonding onto a 2-mil pad would be harder with ultrasonic bonding, because the ultrasonic bond is made on the flat side of the wire, rather than from the top as with a ball bond. This means that the ultrasonic bond spreads over a larger area and creates the possibility of shorting out close geometries such as might be found on a high-frequency transistor. Therefore, the ball size has to be extremely closely controlled. If the machine is misadjusted or the ball size becomes too big, the ball bond will become very large and flat when it is made, possibly shorting out parts of the die metallization.

The thermal compression capillary tip can be used to make bonds very close to discrete components or in deep packages. The back of the standard ultrasonic wedge will hit close components mounted on the substrate or hit the sides of a deep package, in either case making bonding impossible or damaging the components. Special wedges and ultrasonic capillary tips can be used to overcome this. Special ultrasonic wedges with sharply angled wire feeds are available to cover most deep package applications without the use of a capillary.

Ultrasonic bonding is a room-temperature process and avoids the gold-aluminum intermetallic plague problems that are associated with thermal compression bonding. When semiconductor bonding was first started, thermal compression techniques were used almost exclusively. Then it was noted that some of the bonds were failing inexplicably. Upon close examination it was determined that under high-temperature conditions, the gold bonding wire was reacting metallurgically with the aluminum bonding pads on the semiconductor chips. The result was the formation of several gold-aluminum intermetallics at the bonding interface, which made brittle, high-resistance joints, eventually degrading the bonds to the point of failure. The failed joints had discolorations at the failure point, which led to the names "purple plague," "white plague," "black death," and others, depending on the particular intermetallic.

Formation of these intermetallics is time-temperature dependent. The elevated temperatures of 300° to 400°C used during thermal compression bonding are ideal for this, so it is imperative that the bonding cycle be kept very short to prevent plague formation. Aluminum ultrasonic bonding became popular as a room-temperature bonding process, which eliminated the plague-forming high temperatures. Ultrasonic bonding also has the advantage of being an all-aluminum system, eliminating the gold-aluminum interface on the semiconductor surface.

A relatively new technique that is becoming quite popular is pulsed thermal compression bonding. This technique attempts to retain the good features of thermal compression bonding while using the room-temperature processing advantages of ultrasonic bonding.

Pulsed bonding operates by heating the bonding capillary momentarily just when the actual bond is being made. This is accomplished by passing a pulse of electric current through the metal capillary to heat it (by resistance heating) for just an instant. Everything is operated at room temperature until the bond is to be made and then a pulse of heat is applied to make the bond. Immediately after the bond is made, the capillary and bond dissipate the heat and return to room temperature. The substrate-holding fixture or stage may sometimes be operated at low temperature, approximately 100°C, to assist in making the bond.

As well as the plague problem, the high bonding temperatures of thermal compression bonding can lead to electrical degradation of the transistor chips. The last chapter discussed how the temperatures used in eutectic die bonding could degrade the semiconductors. The 300° and 400°C substrate temperatures of thermal compression bonding will continue this degrading process.

Finally, the parameters of ultrasonic bonding, and especially tool pressure, have to be more precisely controlled than those of thermal compression bonding.

Bond Strength

An important feature of wire bonding, as with all of the hybrid manufacturing process, is cleanliness. A substrate or die that has oils on it from being handled, or has been otherwise contaminated, will either be difficult to bond or produce weak bonds. Similarly, any physical pieces of dirt under the bond will weaken it significantly. Bond strengths are measured with special micropull testers. One way of testing bond strength is to use a small spring balance with a hook on it. If no equipment is available, a simple pull test with a pair of tweezers can be used. The bond should feel quite resistant and should come off the bonding pad with a definite "pop". Bonds that slide off the pad too easily are too weak. The pad should be inspected for remains of bonding wire, wire depressions, intermetallics, and other signs of a strong bond. A smooth bonding pad under the bond site indicates a weak bond. One indication of a strong bond is that the bonding wire breaks under a pull test before the bond comes off the pad. Bond strengths of approximately 5 to 8 grams for 1-mil wire are good. Bond strength will depend on the wire specifications, the length of the bond, and the angle of pull. The breaking load for a single bond pulled vertically will be different from that of a wire loop with two bonds, pulled at the middle of the loop. Bond strengths of up to 10 grams or so can be obtained. Of course, lower strengths can be used, but they increase the possibility of bond breakage, especially under conditions such as transfer-molding packaging schemes or when the package is subjected to sudden impacts, for example when dropped.

Wire bond failures cause the highest failure rate of any of the thick film process steps. It is very important to select a bonding scheme and learn everything possible about it. It is also very important to have good, well-trained people manning the bonders.

12 — Microcircuit Packaging

Circuit packaging has been left until the end, because the logical sequence of circuit manufacture calls for the package to be added last. However, the packaging scheme should not be considered last in line when designing and laying out a circuit. The package protects the active portion of the circuit inside and if it does not do its job and protect the circuit, all the careful work in preparing the active electronics inside will be wasted. The packaging scheme is one of the first items to be considered in the circuit design cycle. In many cases the package chosen will determine the size and layout of the circuit.

The ways of packaging a thick film circuit are about as many and as varied as the circuit design itself. It is also one of the most important considerations. The package must protect the circuit itself from physical damage during handling and assembly of the circuit into the final system. It must protect the active elements from the degrading and destructive effects of many storage and operating environments. It provides a way of connecting the active part of the thick film circuit to the outside world through metallic leads that are strong enough to be welded or soldered into the final system. The package helps to dissipate heat produced by the operation of the thick film circuit that would otherwise make the circuit nonoperational. The package is indeed an integral part of the design of the circuit. Figure 12-1 shows a few typical packages.

Most of the packaging schemes used are spin-offs from three basic methods that are the foundations for any packaging scheme:

1. Hermetically sealed and/or commercially made packages.
2. Transfer molding.
3. Potting and casting.

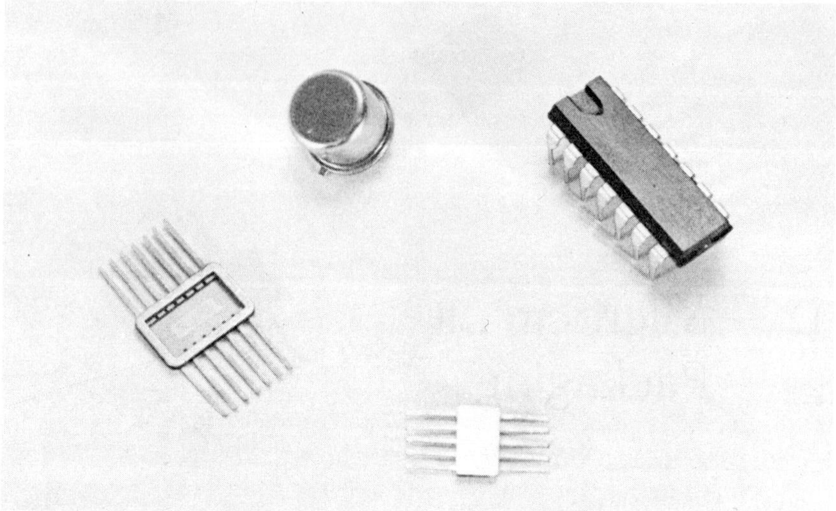

Fig. 12-1. Various packages. From the top, clockwise: TO-5 transistor can; dual-in-line (DIP) plastic molded package; sealed, 10-lead metal flat pack; and unsealed 14-lead hermetic flat pack.

Commercially Made Packages

The most common of the commercially made packages is the flat pack. This consists of a flat, hollow can-like container with insulated leads for external circuit connections coming out through the package walls. The substrate is mounted inside the hollow portion and the wire bonds made to the flat lead endings inside the package. Then a lid is placed on the package and sealed in place, to form a rugged, hermetic assembly. This sequence is diagrammed in Fig. 12-2. Leads are commonly made of Kovar, a cobalt-nickel-iron alloy, coming through a glass seal in the package walls. Kovar is used because it matches the coefficient of expansion of the glass seal very closely. This type of package is designed to be hermetically sealed, so it is very important to thermally match all the package materials to prevent microcracks from developing in the seals and causing leaks to or from the outside. The top and bottom can be made of Kovar or of ceramic. The ceramic is often used when a series of bare chips are to be mounted directly in the package and wired together. The ceramic, which can also have metallized pads and paths on it, is used for an insulating mounting surface.

Flat packs, as in Fig. 12-3(B), are available in sizes from 1/4 x 1/8 inch up to an inch or two square. The number of leads for external connection

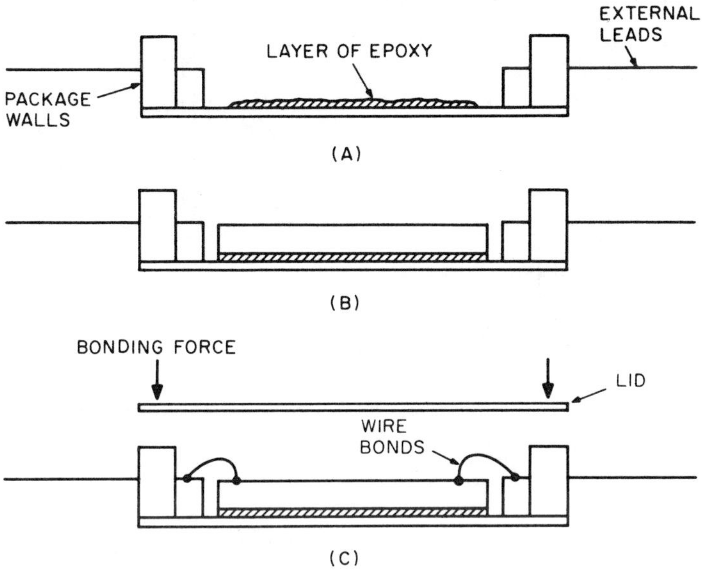

Fig. 12-2. Bonding sequence for putting substrate into a flat pack. (A) A layer of epoxy is spread on the bonding site. (B) The substrate is pressed into the epoxy, and the epoxy is then cured. (C) Wire bonds are made from the circuit to the external leads, and the lid is then sealed on the package.

varics from four or six up to thirty or forty, depending on the package size. Round packages, as well as square and rectangular, are available in various sizes and with various numbers of leads.

Most of the regular transistor cans are available. TO-5 cans, as in Fig. 12-3(A), are available with six, eight, ten, and twelve leads. Also available are TO-8, TO-18, TO-46, TO-47, and others. Standard power transistor packages, such as the TO-3 and TO-66, are available.

All these packages are designed for metal-to-metal seals at the lid surfaces. There are several different methods of sealing these packages. A preform can be used between the lid and the lid-attaching surface. Gold-tin solder preforms with a eutectic temperature of between 260° and 300°C are used. The solder preform is placed between the lid and the package and the solder is reflowed either by a hot plate, infrared heat, a belt oven, or by specially shaped heated tools. Temperature-sensitive epoxy preforms can also be used. The preform is also placed between the lid and the package, then heat is applied to soften the preform and seal the package. The epoxy

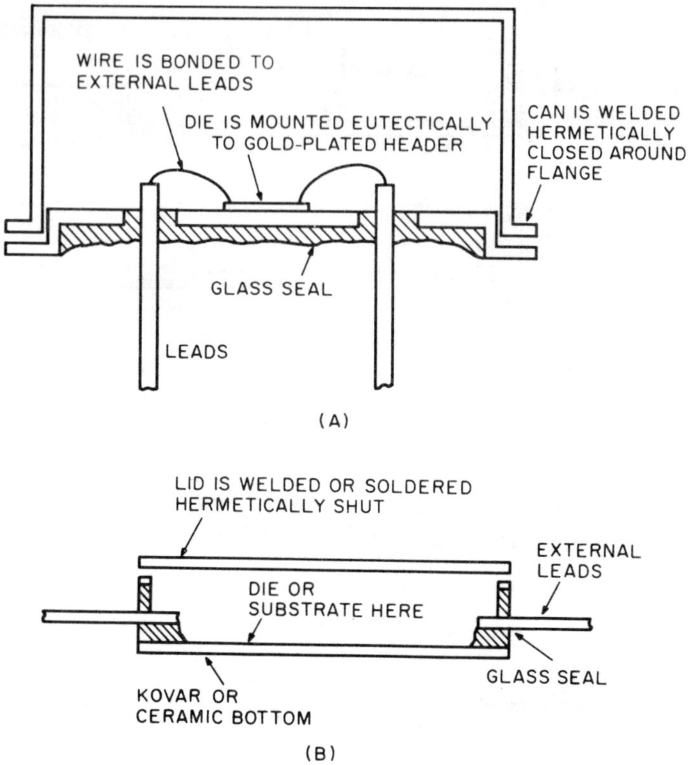

Fig. 12-3. Semiconductor packages. (A) Transistor can. (B) Flat pack.

preforms lower the operating temperature capabilities of the package. The epoxy has a much lower plastic temperature than the solders. As a result, if the circuit is heated up to the plastic temperature during operation, the lid may come off again.

Various welding schemes are also used to seal the package. These all form strong metallurgical seals. Resistance welding, laser welding, and stitch welding heat up the edge of the flat pack or transistor can and form a joint through melting enough of the interface between the lid and the package to form a metallurgical bond. Cold welding, using the effects of pressure alone, can be used.

A consideration for the sealing scheme to be used is the possibility of rework. Packages sealed with solder or epoxy can be opened up again for rework of the circuit or a postmortem of failed circuits. This is accomplished by heating up the circuit again until the solder or epoxy melts, then removing the lid. A package that has been either hot- or cold-welded cannot be reworked. The package has to be destroyed to get into it again. The package-opening process in this case may also damage or destroy the circuit inside.

For a very expensive circuit, such as may be used in MIL-SPEC situations, this rework capability becomes very important in terms of cost.

Many packages are available complete from various manufacturers, and their specification sheets should be consulted for further information. There are also schemes that use modifications of the above ideas. The only limitation is the imagination of the designer. One such scheme uses the substrate as the bottom of the package. Pins for external connection are added either radially out from the substrate or through the substrate and out of the bottom. The active portion of the circuit is covered with a ceramic cup or it is surrounded with glass walls and a Kovar or ceramic lid is added on top. If the substrate is made larger than the actual circuitry and its protective walls, screened conductors can be run under the walls to provide the outside connections. Leads are then soldered or welded onto lead termination pads screened onto the substrate.

Transfer Molding

Plastic-molding of transistors and integrated circuits to make a package is a common practice in the semiconductor industry. This type of packaging is about the cheapest possible. The TO-92 and dual-in-line (DIP) packages are common examples of this. The parts to be encased are placed in a suitable mold and the mold is closed. The molding material in either powder or pellet form is heated to a plastic state, then forced under low pressure into the mold cavity and around the parts to be encapsulated. After the plastic cures, the completed parts are removed.

The tooling costs for transfer molding are quite expensive, so it would not be justified for short-run production. The molding press itself is not cheap and individual molds are required for each size and type of package to be produced. Multicavity molds, used for high-volume, multiple molding are even more expensive. The initial investment is costly compared to other packaging methods, but the cost per piece is extremely low. Over a long run, transfer molding is very inexpensive, as the actual amount and the cost of the material are very low.

The plastics used for transfer molding are thermoset materials. The thermosets cure under the application of heat and cannot be reworked once cured. The most common of the thermosets used for semiconductor molding are epoxy and silicone. Both form strong, rigid packages. The silicone can be removed easily for repair or inspection by the application of various chlorinated solvents, such as trichloroethylene, or the ketone group, such as methyl ethyl ketone. This will cause the silicones to swell and soften. The epoxies can be removed with various proprietary solvents, but are somewhat more difficult to work with and very often destroy the package contents during the epoxy removal process.

There are two possible disadvantages to transfer molding—the heat and the pressure. The circuit assembly must be able to withstand the pressure of the molding compound being forced into the cavity (about 25-150 pounds per square inch) and the molding temperature (about 300°F). A common failure mode is loss of the wire bonds during molding. The pressure of the injected material can be enough to tear the wires off the components. A thin conformal coating of silicone rubber or urethane can be used to give a soft, flexible, buffer coating between the circuit and the molded case.

There is still quite a lot of controversy over the use of plastic semiconductor packages in the electronics industry. The main area of discontent is in the hermeticity. Some people feel that by not having a totally hermetic package, leakage and undesirable chemical contamination of the circuit will occur, with the subsequent failure of the circuit. The argument boils down to a question of whether or not hermeticity is needed. If a slight amount of moisture will affect the circuit operation or the circuit is required to operate in extreme temperature conditions, the circuit is better off in a hermetically sealed metal or glass-based package. For a noncritical commercial application, a plastic package may provide more than enough protection. The other operating drawback of the plastic packages is that they cannot take internally generated heat or high-temperature operating environments. Conversely, the plastic packages cannot dissipate anything like the amount of heat that a ceramic or Kovar package can. But for extremely cheap, rugged packages, transfer molding is an excellent scheme.

Potting and Casting

Potting involves the use of a metal or plastic hollow container in the form of a cup or can. The assembly is positioned inside the container and the bonds made to the external leads. Then the entire assembly is filled with epoxy, silicone, or urethane. The container may or may not have a top to go on the package after the potting material is cured.

Potting is quite economical for short-run, inexpensive production. For long runs, potting is quite time-consuming—transfer molding would be more economical. Potting involves the use (and possibly in-plant manufacture) of the premade shell, plus the extra labor of filling it with potting material. Several materials can be used as the actual potting compound. The common encapsulants are silicone, epoxy, and urethane. Since the overall package would have several shortcomings if these encapsulants were used individually, a combination of materials is usually used. The overall package requirement is that the exterior have a hard, protective, damage-resistant finish. It would seem that an epoxy encapsulant would do the best job, but it has two drawbacks. First, the epoxies in general have a much higher coefficient of expansion than the ceramic substrates and the circuit components. Large temperature changes would result in stresses in the circuit that would crack either the epoxy or the circuit. Either one is undesirable. Second, amine or acid anhydride curing epoxies can release contaminants during curing that react with circuit elements and alter their characteristics. The chemical reduction of palladium-silver resistors is one such effect.

The alternative potting method is to use a soft material such as a silicone rubber as the encapsulating material. This provides a soft, flexible coating that will protect the circuit, yet be flexible enough to allow expansion and a certain degree of flexing of the leads, wires, and components. However, this flexibility is also a disadvantage. Pressure on the package during handling can break the wire bonds in the circuit. The answer is to use a combination of materials. A coating of silicone or urethane is used to provide a soft, flexible buffer coating between the substrate and the circuit elements and the hard outer coating on top. Epoxy can be used for this outer protective coating. The choice of silicone rubber must be made quite carefully. Most silicone rubbers release acetic acid during their cure cycle. This acid will corrode the circuit elements, in particular the metallic portions, and may react chemically to produce other changes. Another possible troublemaker is any ionic contamination present in the coating. This can produce migration and interaction effects with the circuit. Several manufacturers make carefully controlled materials for semiconductor surface coatings.

Casting is basically the same procedure as potting except a mold is used instead of a shell or cup to contain the circuit during encapsulation. The mold is poured full of urethane, epoxy, or other material. This is allowed to harden, the mold is opened, and the cast package is taken out. The mold is then reused for other circuits. Casting, as compared to potting, involves making up the molds to hold the circuits.

A Comparison of Packaging Methods

In conclusion, the packaging scheme used for thick film circuits depends on the intended use of the circuit. Transfer-molded packages require tooling and molding equipment involving a large initial capital investment. This pays off, however, when producing the packages, since the cost per package is very small. The operating and storage environment for plastic packages must be mild, as the package is not hermetic and the casing itself can be damaged by chemical and thermal action.

Commercially available packages are rugged, protect the circuitry well, and can be hermetically sealed. The main drawback of hermetically sealed packages is the price. Since the cost of this type of package can run higher than the actual cost of the electronics inside, it is not a good method to use when considering a low-cost, competitive, commercial market. The decision will depend on the application and the quality desired. The hermetic seal is a desirable feature and is considered essential for high reliability and MIL-SPEC work. The package is purged of all the air in it just before sealing, and is backfilled with nitrogen or some other inert gas. If the package is not purged before sealing, residual gases can interact with the circuitry and degrade the operation of the circuit, especially if operated at high temperatures and over long periods of time. Palladium-silver resistors pose the worst problem, as noted before. Any trace of reducing agents, either a gas or a chemical residue, left in the package will abruptly change the value of the resistors. Another disadvantage with many commercial packages is their size. For small packages, the package can be two or three times the size of the available active area for the substrate. If overall size of the circuit is at a premium, this can be a drawback.

Potting or casting can be a good, economical way to produce small quantities of circuits without the additional costs of tooling for transfer molding. It can be time-consuming, but it is cheaper in small quantities.

13 — Cleanliness

This is one of the last chapters of the text, but it is also one of the most important. If there is any single consideration to be strongly stressed where thick film circuits are concerned, it is cleanliness. The motto of everyone involved with hybrid processing must be "Think Clean." This refers directly to dust and dirt and also to more subtle considerations such as the degrading effects of chemical contamination.

In order to end up with clean circuits the layout and flow of the facility itself must be carefully planned from the beginning. One of the first considerations when planning a hybrid facility is location. Thick film processing equipment is delicate, expensive, and requires some care in handling. Thought must be given to reasonable operating conditions for the equipment and how to keep gross contamination out of the circuits and working area. The facility cannot be set up in the corner of a machine shop or mail room; it is preferable to have an isolated room. Important factors to be considered are construction and physical characteristics of the room, the environmental conditions, the availability of service utilities, and the pattern of personnel flow.

First the room itself must be large enough to allow adequate bench and equipment space for comfortable working conditions. Space should be allotted for expansion and for extra equipment that usually turns out to be necessary after everything has been planned out. Adequate "plumbing" must be figured in. Besides light, heat, and electrical power outlets, the planning has to include compressed air lines, vacuum lines, and nitrogen and water lines, as required. Thick film firing furnaces usually require 220-volt, three-phase electrical lines. Power requirements may range between 5 and 10 KVA for a small furnace, and up to 50 KVA for the larger units. This may require special electrical lines to be run into the area. A sink for cleanup is essential and it is a good idea to include a hand spray and fume hood over the sink for solvent cleaning and any chemical work that may take place.

If screen-stencil-making and photoreduction are to take place within the thick film facility, a separate, enclosed area must be figured in. Normal electronic work facilities these days use fluorescent lighting. Stencil-making will have to be performed in a closed room under tungsten lighting only in order to prevent exposure of the emulsion. Photoreduction will require a complete darkroom facility. Both photographic and stencil work will require running water, sinks, good ventilation, and lighting control. Air conditioning for the whole area should be considered since drying ovens, firing furnaces, and bonders can produce quite a lot of heat. Humidity control is desirable for consistency of operating conditions.

The entire subject of atmosphere control can be as simple or as exotic as the budget will allow. There are some fundamental considerations though. The room air should be dust- and dirt-free and must contain no fumes. Airflow will always be necessary to some extent to remove evaporated solvents and material burned off in the furnace from the immediate work area. Laminar or patterned airflow with inlet and exhaust ducts may be desirable. An important consideration in the airflow scheme is to watch for the direction of the backspray from an air abrasive trimmer. The dust must not be allowed to spray back and contaminate circuits in process or other equipment, such as the screening machine or the paste supply. This may require placing the trimmer in a corner by itself or using a planned airflow to control the backspray. Air abrasive trimmers include vacuum equipment to suck all the dust and debris away from the operator and work area, but these do not usually catch every particle of spray—some usually still leaks out. A fume hood is advisable for using strong solvents such as Freon, acetone, or trichloroethylene.

Having laid out the facility with an eye to airflow and clean working conditions, the processes themselves must be studied for cleanliness and clean working habits.

An important point is cleanliness of the circuits during processing. The most common form of contamination occurs when the substrates are handled. The oils and skin flakes that rub off from the fingers onto the substrates and components can play havoc with many processes.

Starting from the beginning of the process, some of the important points to remember are as follows. The substrates must be clean to begin with. They are not generally suitable for immediate use as they come from the manufacturer, since oils may be present from handling and from machinery, and talc and dust may be in the pores. Either of these conditions can lead to a lowered adhesion strength between the printed films and the substrate. Cleaning can be accomplished by vapor degreasing or by ultrasonic cleaning. An ultrasonic cleaner is very desirable as it will shake out any particles of dirt and talc from the pores. Trichloroethylene in a vapor degreaser or the Freon family of solvents both make good cleaning agents. Good results have also been obtained using detergents and ultrasonic scrubbing. This takes a little longer and is harder to rinse, but can be satisfactorily used. Finally, a rapid run through the furnace at 800° to 1000°C burns off anything that is left and ensures that the substrate is completely dry. Having an unsealed surface, the substrate can absorb moisture. Following this, the clean substrates should be stored in a clean, dust-free environment. Plastic containers with snap lids make excellent containers. Some care must be taken while handling the clean substrates. Carelessly used tweezers can result in metallic scratches on the substrate face. Lint-free gloves, clean rubber gloves, or finger cots are suitable for handling the substrates, and serve to keep the hands from contaminating the surfaces. The same is true all through the processing of the circuits. Handle only with tweezers or gloves. Skin oils on a conductor make wire bonding impossible or, at best, erratic. Skin salts from perspiration cause corrosion and other metallurgical effects.

Pastes should not come into contact with the hands for two reasons: the paste becomes contaminated with dead skin flakes and skin oils and dirt; and the paste is toxic.

Firing atmosphere is important and, in the case of palladium-silver resistors, it is critical. Reducing agents must be avoided and humidity must be controlled for best results. Strong vapors in the firing area must be kept out of the furnace for consistent and good results.

Care must be taken with the circuits during the active device-processing steps. To begin with, the semiconductor dice themselves must be clean. If they are purchased bulk-packed, the motion of the dice within the vial will cover them with silicon dust and debris. Wax and dust may be left on them from the scribing and breaking step at the manufacturer's. The dice can be cleaned off with Freon or trichloroethylene.

Two serious problems can occur with semiconductor dice. One is chemical contamination from the environment leading to device degradation. The other is oxidation of the aluminum metallizations interfering with wire bonding. With ultrasonic wire bonding this is not too serious a problem and a small amount of oxidation will assist the abrasive scrubbing action. But with thermal compression bonding, these oxides can result in a weak bond or make bonding impossible. The solution to both these problems is to store the dice in an inert atmosphere. Small storage cabinets with self-sealing doors are available for this purpose. The entire cabinet can be filled with nitrogen to prevent oxidation or contamination. During hot bonding steps, any contamination on the dice can become burned in. During wire bonding, these contaminants can make bonding weak, erratic, or impossible.

If soldering steps are used as part of the circuit assembly, the type of soldering flux must be chosen with care. Chloride or acid fluxes are taboo because of the corrosive effects on both the conductors and the resistors, and chemical reaction problems with the resistors. The flux can be a mild organic type of flux and should be cleaned off thoroughly after soldering is complete. Again, vapor degreasing or ultrasonic cleaning with a suitable solvent is recommended.

Finally, the finished circuit must be clean before encapsulation or sealing. No residues that could have long-term degrading or corrosive effects must be left on the circuit.

If packages are hermetically sealed, the sealing atmosphere is very important. The normal air in the working area can contain fumes and particles that can interact with the circuit elements, and can cause problems later on if sealed into the package. These effects usually show up after long storage and especially at elevated temperatures. The generally used solution is to cover the sealing area and flush out the package being sealed at the same time with a blanket of inert gas such as clean, dry nitrogen. This ensures that an inert atmosphere is sealed into the package.

Throughout the entire manufacturing process care must be taken to exclude gross contaminants such as dust, lint, dirt, hair, and other similar factors. These can cause immediate physical circuit problems or, worse yet, the interaction effects may be more subtle and not show up until the units are finished and out into the field.

14 — Safety

The hybrid circuit manufacturing process uses many chemicals and materials that are toxic, highly flammable, and otherwise hazardous. Some of the individual process steps are themselves potentially dangerous. Problems similar to these exist in every manufacturing operation, but if they are recognized by the people involved, the materials and processes can be handled with a minimum amount of danger to the operating personnel. It is important to educate people in correct handling methods and precautionary measures. Constant contact with these hazards often leads to carelessness and a situation ripe for an accident. It is important to periodically remind personnel of these hazards to prevent a potential accident situation from occurring.

This chapter will present some of the potential hazards that exist in a thick film facility, and try to point out some of the ways that these problems can be controlled. Obviously, not all the problems can be mentioned. They will depend a great deal on the individual manufacturing process. So it is up to the individual circuit manufacturers to find out the potential problems associated with their processes and materials.

Optical Hazards

There are two optical problems that may be commonly encountered in thick film work. One is ultraviolet light sources and the other is laser beams. Both of these can cause permanent blindness.

Ultraviolet light is used commonly to expose direct and indirect emulsions to produce screen stencil images. Ultraviolet light can be generated by several different sources. Fluorescent tubes, mercury vapor lamps, "sun-ray" lamps, carbon-arc lamps, and quartz-iodine lamps are among the most common. Here a distinction must be made. These sources generate two types of ultraviolet radiation—longwave and shortwave. Longwave ultraviolet radiation is produced by black light fluorescent tubes. It is not generally considered dangerous, although it is advisable to wear glasses with protective filters for extended periods of viewing. These can be obtained from companies that specialize in optical filters. Shortwave ultraviolet radiation is the dangerous one, and suitable eye protection must always be worn when working with it. Shortwave radiation is generated by mercury vapor lamps and carbon-arc lamps. A person without suitable protective filters should never look directly at a shortwave ultraviolet source. The effect is similar to looking directly at the sun. The radiation will produce severe burns on the retina at the back of the eyes. These burns produce gray or black "dead" areas in the field of vision that are nonregenerative and nonrepairable. Prolonged viewing without protection will lead to total blindness. Protective goggles should always be worn and, if possible, the work should be performed in an enclosed box with suitable interlocks to prevent accidental exposure of any personnel.

The second hazard area is in the use of lasers, which are becoming more common for resistor trimming, substrate cutting, and other uses. The safety protection needed depends on the type of laser and the system it is built for. The manufacturer's recommendations should always be followed.

Carbon dioxide lasers operate at 10.6 microns in the far infrared region. Used for hybrid work, they are low-power lasers that are considered to be low-hazard devices. The long wavelength of the CO_2 laser makes ordinary glass opaque to the beam. This means that ordinary eyeglasses or microscope optics can be used as safety glasses. The human eye is theoretically opaque to the CO_2 wavelength, so the possibility of eye damage is very remote. However, it is wise to be cautious and wear protective glasses.

The dangerous lasers are the YAG and Argon lasers. They should never be viewed directly without suitable eye protection since they will burn the retina, causing partial or total loss of vision similar to that caused by the carbon arc lamp. Some laser trimmers use enclosed filter hoods with safety interlocks to shield the operating personnel; others use closed circuit television with remote controls for operating. None of the laser safety devices should ever be altered to defeat the protective features.

Process Problems

Air abrasive trimming is a messy operation at best, but worse than this, it can be a respiratory hazard if the abrasive dust is not controlled. A vacuum system surrounding the trim area is generally enough to remove the dust and overspray that occurs. Special precautions must be taken to ensure that all the dust is removed when trimming beryllia (beryllium oxide) substrates. The dust must also be very closely controlled when drilling or machining beryllia.

Inhalation of beryllia dust causes a lung disease similar to silicosis, contracted by miners after inhaling coal and rock dust for long periods of time.

Ultrasonic cleaning is becoming commonplace in many industrial situations. There is the possibility of tissue damage to parts of the body which repeatedly come into contact with a high-power ultrasonic cleaner or cleaning solution. It is most probable that physical sensations, such as burning, would cause the affected area to be withdrawn before actual damage could occur. It is only wise not to work in the tank while the machine is in operation. Another hazard with high-power ultrasonics is the possibility of hearing damage under certain conditions. Sound-level checks can be made to ensure safe working levels.

Chemicals

The sensitizers used for emulsion preparation in thick film screenmaking are strong irritants. The sensitizers commonly used are water-based solutions of ammonium dichromate, potassium dichromate, or sodium dichromate. All the dichromates can cause serious skin and lung irritations and ulcers that are painful and slow to heal. When handling any of the dichromates, especially in powder form, it is important to wear rubber gloves and safety glasses and to avoid any skin contact with the dust. Good housekeeping procedures should be followed in any chemical work.

Most photographic processing chemicals can cause skin and respiratory irritation, if frequent or prolonged contact takes place. This can again be prevented by wearing rubber gloves and by providing a good ventilation system for the darkroom.

Most of the heavy metals are poisonous and, even worse, are cumulative poisons. The body excretes them slowly or not at all. Thallium is one such metal. Thallium sulfate is used as a pest-control poison. Thallium may be encountered when using thallium-based thick film pastes. Thallium compounds in solution are readily absorbed through the skin and can be a hazard to all the body organs. However they can be used with perfect safety, provided suitable precautions are taken to prevent any skin contact.

Many of the RTV silicone rubbers release acetic acid during their curing cycle. This will irritate the eyes and respiratory system if the fumes are trapped in the work area. A good ventilation system will eliminate this problem.

Epoxy resins of various types are in widespread use as adhesives and potting materials. Epoxy resins and curing agents will cause dermatitis and various skin irritations upon repeated skin contact. As with other chemical formulations, the best way to avoid these problems is to avoid all skin contact by wearing gloves. Disposable polyethylene gloves that can be used once and then thrown away are ideal for mixing and using epoxies. Good ventilation to prevent the accumulation of fumes in the work area is also a good practice. If skin contact does occur, the area should be washed off at once with alcohol and then washed well with soap and water.

The pastes used for thick film printing use organic solvents for viscosity control. These solvents are driven off after printing to leave a dried pattern before firing. The pastes should not be allowed to come into contact with the skin, as solvents can cause various problems. Solvents will be discussed further on. As a good housekeeping practice, ventilation is required in the printing and drying area to prevent a buildup of solvent vapors and the resulting toxicity and fire or explosion hazards.

The common reagent acids such as hydrochloric, nitric, and sulfuric acids may be used. They should be handled with standard laboratory precautions. Safety glasses, protective gloves, and strong ventilation are recommended. Remember to add acids to water when diluting them and never add water to acids. This can cause overheating and splattering of the acid.

One acid that may be encountered that is not as well known as the others is hydrofluoric acid (HF). It is not a chemically strong acid like nitric or sulfuric acid, but it is probably one of the most dangerous to use. The strong reagent acids, if spilled on the skin, will cause immediate burns that can be felt and treated immediately. Hydrofluoric acid does not cause an immediate burning sensation and may go unnoticed for some time until quite serious burns are sustained. The burns that are caused are painful and slow to heal. Hydrofluoric acid is used mainly for glass etching, but is also encountered in glaze or ceramic work.

Compressed Gases

Normal precautions should be taken when dealing with any gas. Use common sense. Cylinders should be stored in a cool, dry place and should not be subjected to any violent mechanical handling. Cylinders should have safety straps or chains to prevent them from being knocked over. Regulators, lines, and fittings should all be treated with care and not mechanically abused. Pressure ratings of hoses and other devices should be observed.

Nitrogen is an inert gas and does not usually provide any chemical reaction problem. It is encountered in die and wire bonding machinery and package-sealing equipment as an inert cover gas. It may also be used in furnaces and ovens as a cover gas. The only problem that may be encountered occurs when large amounts of nitrogen are released to the work area or when a leak occurs in the system. Nitrogen can be breathed, but it does not support life. In large quantities, it prevents an adequate supply of oxygen from reaching the lungs, leading to eventual death by suffocation. This can be controlled by preventing leaks and having good, fresh-air ventilation.

Hydrogen is used most commonly in thermal compression wire bonding equipment to provide a flame for ball-forming torches. It is, naturally, a highly flammable gas and should be kept away from heat or open flame, and preferably from other flammable materials. Special care should be taken to see that there are no leaks anywhere in the gas line or fittings. Explosions can occur with the right mixture of hydrogen and air.

Oxygen is sometimes used to enrich the atmosphere in various processes. Oxygen must not be stored with any combustible material. Oil and grease must never be used on oxygen-handling equipment or lines, and no flammable material can be allowed to come into contact with oxygen systems. Oxygen is not flammable itself, but it will support combustion and increase the burning rate tremendously.

Solvents

Solvents of various kinds are very commonly encountered at various steps in the thick film process. Most solvents are toxic to some degree. This section will try to point out some of the key features of solvents that may be commonly encountered.

The primary problem with solvents is that, in contact with the skin, the normal solvent action will remove the natural skin oils and fats, leading to local skin irritation and dermatitis. Many solvents are toxic to the human system, attacking the liver, kidneys, nervous system, and respiratory tract. Some solvents are absorbed directly through the unbroken skin into the system. The best prevention is not to allow any skin contact at all, by using solvent-resistant gloves and safety glasses or a face shield.

Acetone, although of low toxicity, is very volatile and highly flammable. It can easily form explosive mixtures with air, can irritate the eyes and respiratory tract, and will cause skin irritation upon prolonged contact.

Butyl carbitol acetate is a flammable liquid and a moderate fire hazard. Wear goggles since it is damaging to the eyes.

Benzene is one of the most dangerous solvents to use. It is a cumulative poison that is absorbed directly through the skin as well as inhaled. It is highly volatile and flammable and is therefore a dangerous fire and explosion hazard.

Butyl cellosolve is highly toxic and a moderate fire hazard. It is irritating to the eyes, skin, and respiratory tract, and also causes kidney inflammation.

Carbon tetrachloride is probably the most dangerous solvent to use and is not recommended. It is extremely toxic and is absorbed through unbroken skin, causing cumulative damage to the liver, kidneys, and nervous system. It is not flammable, but when subjected to high temperatures it releases hydrochloric acid and phosgene (carbon oxychloride). Phosgene is highly corrosive and poisonous and affects the throat, eyes, and lungs.

Ethyl alcohol is not generally considered toxic, but will irritate the respiratory tract and cause headaches.

Freon TF (Du Pont) is nonflammable and has very low toxicity. The greatest danger is that of suffocation. Good ventilation will take care of this. Prolonged contact will dry the skin of natural oils, so rubber gloves should be worn for protection. When subjected to high temperatures, it can form toxic byproducts such as hydrogen chloride, hydrogen fluoride, and phosgene.

Isopropyl alcohol is a dangerous fire hazard and flammable as a liquid. Although not considered very toxic, it will irritate eyes, nose, and throat.

Toluene is a highly flammable liquid. Although moderately toxic, it is absorbed through the skin and its toxic effects are cumulative. Irritating to the skin (through defatting action) and especially to the eyes and respiratory tract. All skin contact should be avoided.

Trichloroethylene is nonflammable and nonexplosive at room temperatures, but high temperature or an open flame will cause decomposition into toxic and corrosive byproducts, such as hydrogen chloride and phosgene. It is irritating to the nose, throat, eyes, and skin and may cause kidney damage. It is toxic when inhaled or absorbed through the skin. All personal contact with it should be avoided. Good ventilation, rubber gloves, and eye protection are good precautions.

Xylene is highly toxic and a great fire and explosion hazard under some conditions. It is absorbed through the skin.

Certain things can be said about solvents in general. They tend to defat the skin by their natural solvent action, leading to skin irritation of various types. All are toxic to some degree, either by inhalation or direct absorption through the unbroken skin. Most will irritate the eyes and respiratory tract. Many form toxic byproducts upon heating. Many are highly flammable.

The best place for solvents and solvent work is in a fume hood with good ventilation to the outside, or some other area with good, fresh-air ventilation. Rubber gloves should be worn to prevent any skin contact, and safety glasses or a face shield should be worn to protect the eyes. The eyes, in particular, are very sensitive and easily damaged. Most chemicals are absorbed into the system very rapidly through the eyes. If eye contact or body contact over a large area should occur, flush well with running water and call a doctor, no matter how insignificant the injury may seem.

With all chemicals, read the labels carefully, and be sure to follow the manufacturers' instructions or handling precautions.

As mentioned at the beginning of this chapter, a thick film facility has potential hazards associated with it. Therefore it cannot be overemphasized that all the people who may come into contact with these materials and process steps must be educated as to the potential dangers. This must be followed up by constant monitoring by supervisory personnel to make sure that material- and equipment-handling rules are not being violated. It is easy to become careless and overconfident after working with a potentially dangerous material and not having any accidents. If this attitude creeps in, a serious mishap is more likely to occur. It is far better to wear safety glasses a hundred times, than to wear them ninety-nine times and have acid splash in the eyes on the hundredth.

Constant vigilance and a careful attitude towards these hazards can prevent mishaps and injury.

Bibliography

American Lava Corporation: *Mechanical and Electrical Properties of AlSiMag Ceramics.* Chart #691.

Apollo Lasers, Inc.: *Lasertrim Users Manual.* Technical Bulletin, 1970.

Alfred T. Batch: *Surveying Current Microelectronic Packaging.* Electronic Packaging and Production, March 1971.

J. Howard Beck: *Firing Thick Film Integrated Circuits.* First Technical Thick Film Symposium, Palo Alto, California, 1967.

Rufus Benton: *Industrial Hazards in the Chemical Machining Industry.* PCM-PCE, April 1970.

E.W. Callander and L. Merrill Palmer: *Care and Feeding of Semiconductor Chips.* Electronic Products, July 1970.

Sigmund Cohn Corporation: *Gold Bonding Wires.* Technical Bulletin, 1967.

Coors Porcelain Company: *Alumina Ceramic Substrates.* Bulletin #7012.

Lewis H. Coronis: *Indirect and Direct Etched Metal Masks for Deposition Control and Fine Line Printing.* 1969 Hybrid Microelectronics Symposium, Dallas, Texas.

Howard K. Dicken: *Surveying Chip Interconnection Techniques.* Electronic Packaging and Production, October 1970.

EDN Editorial Staff: *Minicomponents – Building Blocks for Hybrids.* EDN, September 1970.

EDN Staff: *Hybrids Today.* EDN, October 1970.

E.I. du Pont de Nemours & Co., Inc.: *The Screen Printing Process for Thick Film Microcircuit Preparation.* Technical Bulletin #A-53809, 1967.

E.I. du Pont de Nemours & Co., Inc.: *Handling Fluorocarbon Solvents Safely.* Technical Bulletin S-30, 1969.

Electronic Packaging and Production Staff Report: *Ceramics for Electronic Packaging.* Chicago, Illinois.

ElectroMaterials Corporation of America: *FIRON Pastes for Thick Film Microcircuits.* EMCA Bulletin #125.

Gerald K. Fehr: *Microcircuit Packaging and Assembly – State of the Art.* Solid State Technology, August 1970.

George Flynn: *Survey of Chip Resistors.* Electronic Products, January 18, 1971.

F.Z. Keister and D. Auda: *Design Guide for Thick Film Hybrid Microcircuits.* Second Symposium on Hybrid Microelectronics, Boston, Massachusetts, 1967.

Albert Kosloff: *Screen Printing Electronic Circuits.* The Signs of the Times Publishing Company, Cincinnati, Ohio, 1968.

A. Koudounaris: *The Case for Cold Welding.* Electronic Packaging and Production, March 1969.

W.H. Legat, W.C. Rosvold, and W.B. Ward: *Beam Lead Devices – The New Hybrid Circuit Element.* 1968 Hybrid Microelectronics Symposium.

David E. Maguire: *Tantalum Chip Capacitors.* 1968 Hybrid Microelectronics Symposium.

Howard W. Markstein: *Passive Microdiscrete Components.* Electronic Packaging and Production, December 1969.

Clare E. Nordquist: *Design Criteria for Ceramic Substrates.* 1968 Hybrid Microelectronics Symposium.

Sandia Laboratories: *Contamination Control Handbook.* NASA SP-5076, 1969.

San Fernando Electric Manufacturing Company: *Ceramic Chip Capacitor Handbook.* San Fernando, California, 1970.

N. Irving Sax: *Handbook of Dangerous Materials.* Reinhold Publishing Corporation, New York, New York, 1951.

Irwin Sherry: *Chip Capacitors.* Electronic Products, March 15, 1970.

Sidney C. Silver: *Components and Chips for ICs.* Electronic Products, March 15, 1969.

Donald J. Spigarelli: *Choosing a Multiple-Zone Firing Furnace for Thick Film Production.* Technical Bulletin, Watkins-Johnson Company, Santa Cruz, California, 1967.

Larry Steckler: *Hybrid IC Packages.* Electro-Technology, July 1968.

Sidney J. Stein and Louis Ugol: *Effect of Firing Conditions on Stability and Properties of Glaze Resistors.* Technical Bulletin, Electro-Science Laboratories, Philadelphia, Pennsylvania, 1968.

Stephen A. Thompson: *Hybrid Resistor Trimming: An Industry Report.* The Electronic Engineer, October 1970.

Paul R. Van Loan and Charles F. Parks: *Thallium Oxide Glaze Resistors on Alumina with Silver Terminations.* Technical Bulletin, Airco Speer Research Laboratories.

Lawrence H. Van Vlack: *Physical Ceramics for Engineers.* Addison-Wesley Publishing Company, Inc., Reading, Massachusetts, 1964.

R.L. Walters and M.J. Weiner: *Resistor Trimming and Other Micromachining with a YAG Laser.* Pennsylvania State University Engineering Seminar, 1969.

Raymond M. Warner, Jr. and James N. Fordemwalt: *Integrated Circuits: Design Principles and Fabrication.* McGraw-Hill Book Company, New York, New York, 1965.

Appendix: Further Reading

The following sources are recommended for further reading and reference on thick film microcircuitry:

1. For inspection criteria, these two sources may be helpful.

   a. *Test Methods and Procedures for Microelectronics.*
   MIL-STANDARD 883.
   Department of Defense publication.

   b. *Dice and Wafer Visual Inspection Criteria for Discrete Transistors.*
   National Semiconductor Corporation, Danbury, Connecticut.

2. Useful sources of information are the ISHM and NEPCON seminar proceedings.

   a. International Society for Hybrid Microelectronics
   1410 Higgins Road
   Parkridge, Illinois 60068

   b. National Electronic Packaging and Production Conference
   Industrial and Scientific Conference Management, Inc.
   222 W. Adams Street
   Chicago, Illinois 60606

Index

Index